Rs. 395/-

Springer Finance

Springer Finance

Springer Finance is a programme of books aimed at students, academics and practitioners working on increasingly technical approaches to the analysis of financial markets. It aims to cover a variety of topics, not only mathematical finance but foreign exchanges, term structure, risk management, portfolio theory, equity derivatives, and financial economics.

M. Ammann, Credit Risk Valuation: Methods, Models, and Application (2001)

K. Back, A Course in Derivative Securities: Introduction to Theory and Computation (2005)

E. Barucci, Financial Markets Theory. Equilibrium, Efficiency and Information (2003)

T.R. Bielecki and M. Rutkowski, Credit Risk: Modeling, Valuation and Hedging (2002)

N.H. Bingham and R. Kiesel, Risk-Neutral Valuation: Pricing and Hedging of Financial Derivatives (1998, 2nd ed. 2004)

D. Brigo and F. Mercurio, Interest Rate Models: Theory and Practice (2001)

R. Buff, Uncertain Volatility Models-Theory and Application (2002)

R.A. Dana and M. Jeanblanc, Financial Markets in Continuous Time (2002)

G. Deboeck and T. Kohonen (Editors), Visual Explorations in Finance with Self-Organizing Maps (1998)

R.J. Elliott and P.E. Kopp, Mathematics of Financial Markets (1999, 2nd ed. 2005)

H. Geman, D. Madan, S.R. Pliska and T. Vorst (Editors), Mathematical Finance-Bachelier Congress 2000 (2001)

M. Gundlach, F. Lehrbass (Editors), CreditRisk$^+$ in the Banking Industry (2004)

B.P. Kellerhals, Asset Pricing (2004)

Y.-K. Kwok, Mathematical Models of Financial Derivatives (1998)

M. Külpmann, Irrational Exuberance Reconsidered (2004)

P. Malliavin and A. Thalmaier, Stochastic Calculus of Variations in Mathematical Finance (2005)

A. Meucci, Risk and Asset Allocation (2005)

A. Pelsser, Efficient Methods for Valuing Interest Rate Derivatives (2000)

J.-L. Prigent, Weak Convergence of Financial Markets (2003)

B. Schmid, Credit Risk Pricing Models (2004)

S.E. Shreve, Stochastic Calculus for Finance I (2004)

S.E. Shreve, Stochastic Calculus for Finance II (2004)

M. Yor, Exponential Functionals of Brownian Motion and Related Processes (2001)

R. Zagst, Interest-Rate Management (2002)

Y.-L. Zhu, X. Wu, I.-L. Chern, Derivative Securities and Difference Methods (2004)

A. Ziegler, Incomplete Information and Heterogeneous Beliefs in Continuous-time Finance (2003)

A. Ziegler, A Game Theory Analysis of Options (2004)

Steven E. Shreve

Stochastic Calculus for Finance I

The Binomial Asset Pricing Model

Springer

Steven E. Shreve
Department of Mathematical Sciences
Carnegie Mellon University
Pittsburgh, PA 15213
USA
shreve@cmu.edu

Mathematics Subject Classification (2000): 60-01, 60H10, 60J65, 91B28

Library of Congress Cataloging-in-Publication Data
Shreve, Steven E.
Stochastic calculus for finance / Steven E. Shreve
p. cm.-(Springer finance series)
Includes bibliographical references and index.
Contents v. 1. The binomial asset pricing model.
ISBN 0-387-40100-8
1. Finance—Mathematical models—Textbooks. 2. Stochastic analysis—Textbooks.
I.Title. II. Springer finance.
HG106.S57 2003
332'.01'51922—dc22 2003063342

ISBN 978-0-387-40100-3
ISBN 0-387-40100-8

springer.com

Shreve: Stochastic Calculus for Finance I

First Indian Reprint 2009

ISBN 978-81-8489-272-7

This edition is published by Springer (India) Private Limited, (a part of Springer Science+Business Media), Registered Office: 3rd Floor, Gandharva Mahavidyalaya, 212, Deen Dayal Upadhyaya Marg, New Delhi – 110 002, India.

Printed in India by Rakmo Press (P) Ltd., New Delhi.

To my students

Preface

Origin of This Text

This text has evolved from mathematics courses in the Master of Science in Computational Finance (MSCF) program at Carnegie Mellon University. The content of this book has been used successfully with students whose mathematics background consists of calculus and calculus-based probability. The text gives precise statements of results, plausibility arguments, and even some proofs, but more importantly, intuitive explanations developed and refined through classroom experience with this material are provided. Exercises conclude every chapter. Some of these extend the theory and others are drawn from practical problems in quantitative finance.

The first three chapters of Volume I have been used in a half-semester course in the MSCF program. The full Volume I has been used in a full-semester course in the Carnegie Mellon Bachelor's program in Computational Finance. Volume II was developed to support three half-semester courses in the MSCF program.

Dedication

Since its inception in 1994, the Carnegie Mellon Master's program in Computational Finance has graduated hundreds of students. These people, who have come from a variety of educational and professional backgrounds, have been a joy to teach. They have been eager to learn, asking questions that stimulated thinking, working hard to understand the material both theoretically and practically, and often requesting the inclusion of additional topics. Many came from the finance industry, and were gracious in sharing their knowledge in ways that enhanced the classroom experience for all.

This text and my own store of knowledge have benefited greatly from interactions with the MSCF students, and I continue to learn from the MSCF alumni. I take this opportunity to express gratitude to these students and former students by dedicating this work to them.

Acknowledgments

Conversations with several people, including my colleagues David Heath and Dmitry Kramkov, have influenced this text. Łukasz Kruk read much of the manuscript and provided numerous comments and corrections. Other students and faculty have pointed out errors in and suggested improvements of earlier drafts of this work. Some of these are Jonathan Anderson, Bogdan Doytchinov, Steven Gillispie, Sean Jones, Anatoli Karolik, Andrzej Krauze, Petr Luksan, Sergey Myagchilov, Nicki Rasmussen, Isaac Sonin, Massimo Tassan-Solet, David Whitaker and Uwe Wystup. In some cases, users of these earlier drafts have suggested exercises or examples, and their contributions are acknowledged at appropriate points in the text. To all those who aided in the development of this text, I am most grateful.

During the creation of this text, the author was partially supported by the National Science Foundation under grants DMS-9802464, DMS-0103814, and DMS-0139911. Any opinions, findings, and conclusions or recommendations expressed in this material are those of the author and do not necessarily reflect the views of the National Science Foundation.

Pittsburgh, Pennsylvania, USA
December 2003

Steven E. Shreve

Contents

Introduction

Background

By awarding Harry Markowitz, William Sharpe, and Merton Miller the 1990 Nobel Prize in Economics, the Nobel Prize Committee brought to worldwide attention the fact that the previous forty years had seen the emergence of a new scientific discipline, the "theory of finance." This theory attempts to understand how financial markets work, how to make them more efficient, and how they should be regulated. It explains and enhances the important role these markets play in capital allocation and risk reduction to facilitate economic activity. Without losing its application to practical aspects of trading and regulation, the theory of finance has become increasingly mathematical, to the point that problems in finance are now driving research in mathematics.

Harry Markowitz's 1952 Ph.D. thesis *Portfolio Selection* laid the groundwork for the mathematical theory of finance. Markowitz developed a notion of mean return and covariances for common stocks that allowed him to quantify the concept of "diversification" in a market. He showed how to compute the mean return and variance for a given portfolio and argued that investors should hold only those portfolios whose variance is minimal among all portfolios with a given mean return. Although the language of finance now involves stochastic (Itô) calculus, management of risk in a quantifiable manner is the underlying theme of the modern theory and practice of quantitative finance.

In 1969, Robert Merton introduced stochastic calculus into the study of finance. Merton was motivated by the desire to understand how prices are set in financial markets, which is the classical economics question of "equilibrium," and in later papers he used the machinery of stochastic calculus to begin investigation of this issue.

At the same time as Merton's work and with Merton's assistance, Fischer Black and Myron Scholes were developing their celebrated option pricing formula. This work won the 1997 Nobel Prize in Economics. It provided a satisfying solution to an important practical problem, that of finding a fair price for a European call option (i.e., the right to buy one share of a given stock at a specified price and time). In the period 1979–1983, Harrison, Kreps, and Pliska used the general theory of continuous-time stochastic processes to put the Black-Scholes option-pricing formula on a solid theoretical basis, and, as a result, showed how to price numerous other "derivative" securities.

Many of the theoretical developments in finance have found immediate application in financial markets. To understand how they are applied, we digress for a moment on the role of financial institutions. A principal function of a nation's financial institutions is to act as a risk-reducing intermediary among customers engaged in production. For example, the insurance industry pools premiums of many customers and must pay off only the few who actually incur losses. But risk arises in situations for which pooled-premium insurance is unavailable. For instance, as a hedge against higher fuel costs, an airline may want to buy a security whose value will rise if oil prices rise. But who wants to sell such a security? The role of a financial institution is to design such a security, determine a "fair" price for it, and sell it to airlines. The security thus sold is usually "derivative" (i.e., its value is based on the value of other, identified securities). "Fair" in this context means that the financial institution earns just enough from selling the security to enable it to trade in other securities whose relation with oil prices is such that, if oil prices do indeed rise, the firm can pay off its increased obligation to the airlines. An "efficient" market is one in which risk-hedging securities are widely available at "fair" prices.

The Black-Scholes option pricing formula provided, for the first time, a theoretical method of fairly pricing a risk-hedging security. If an investment bank offers a derivative security at a price that is higher than "fair," it may be underbid. If it offers the security at less than the "fair" price, it runs the risk of substantial loss. This makes the bank reluctant to offer many of the derivative securities that would contribute to market efficiency. In particular, the bank only wants to offer derivative securities whose "fair" price can be determined in advance. Furthermore, if the bank sells such a security, it must then address the hedging problem: how should it manage the risk associated with its new position? The mathematical theory growing out of the Black-Scholes option pricing formula provides solutions for both the pricing and hedging problems. It thus has enabled the creation of a host of specialized derivative securities. This theory is the subject of this text.

Relationship between Volumes I and II

Volume II treats the continuous-time theory of stochastic calculus within the context of finance applications. The presentation of this theory is the raison d'être of this work. Volume II includes a self-contained treatment of the probability theory needed for stochastic calculus, including Brownian motion and its properties.

Volume I presents many of the same finance applications, but within the simpler context of the discrete-time binomial model. It prepares the reader for Volume II by treating several fundamental concepts, including martingales, Markov processes, change of measure and risk-neutral pricing in this less technical setting. However, Volume II has a self-contained treatment of these topics, and strictly speaking, it is not necessary to read Volume I before reading Volume II. It is helpful in that the difficult concepts of Volume II are first seen in a simpler context in Volume I.

In the Carnegie Mellon Master's program in Computational Finance, the course based on Volume I is a prerequisite for the courses based on Volume II. However, graduate students in computer science, finance, mathematics, physics and statistics frequently take the courses based on Volume II without first taking the course based on Volume I.

The reader who begins with Volume II may use Volume I as a reference. As several concepts are presented in Volume II, reference is made to the analogous concepts in Volume I. The reader can at that point choose to read only Volume II or to refer to Volume I for a discussion of the concept at hand in a more transparent setting.

Summary of Volume I

Volume I presents the binomial asset pricing model. Although this model is interesting in its own right, and is often the paradigm of practice, here it is used primarily as a vehicle for introducing in a simple setting the concepts needed for the continuous-time theory of Volume II.

Chapter 1, *The Binomial No-Arbitrage Pricing Model*, presents the no-arbitrage method of option pricing in a binomial model. The mathematics is simple, but the profound concept of risk-neutral pricing introduced here is not. Chapter 2, *Probability Theory on Coin Toss Space*, formalizes the results of Chapter 1, using the notions of martingales and Markov processes. This chapter culminates with the risk-neutral pricing formula for European derivative securities. The tools used to derive this formula are not really required for the derivation in the binomial model, but we need these concepts in Volume II and therefore develop them in the simpler discrete-time setting of Volume I. Chapter 3, *State Prices*, discusses the change of measure associated with risk-neutral pricing of European derivative securities, again as a warm-up exercise for change of measure in continuous-time models. An interesting application developed here is to solve the problem of optimal (in the sense of expected utility maximization) investment in a binomial model. The ideas of Chapters 1 to 3 are essential to understanding the methodology of modern quantitative finance. They are developed again in Chapters 4 and 5 of Volume II.

The remaining three chapters of Volume I treat more specialized concepts. Chapter 4, *American Derivative Securities*, considers derivative securities whose owner can choose the exercise time. This topic is revisited in a continuous-time context in Chapter 8 of Volume II. Chapter 5, *Random Walk*, explains the reflection principle for random walk. The analogous reflection principle for Brownian motion plays a prominent role in the derivation of pricing formulas for exotic options in Chapter 7 of Volume II. Finally, Chapter 6, *Interest-Rate-Dependent Assets*, considers models with random interest rates, examining the difference between forward and futures prices and introducing the concept of a forward measure. Forward and futures prices reappear at the end of Chapter 5 of Volume II. Forward measures for continuous-time models are developed in Chapter 9 of Volume II and used to create forward LIBOR models for interest rate movements in Chapter 10 of Volume II.

Summary of Volume II

Chapter 1, *General Probability Theory*, and Chapter 2, *Information and Conditioning*, of Volume II lay the measure-theoretic foundation for probability theory required for a treatment of continuous-time models. Chapter 1 presents probability spaces, Lebesgue integrals, and change of measure. Independence, conditional expectations, and properties of conditional expectations are introduced in Chapter 2. These chapters are used extensively throughout the text, but some readers, especially those with exposure to probability theory, may choose to skip this material at the outset, referring to it as needed.

Chapter 3, *Brownian Motion*, introduces Brownian motion and its properties. The most important of these for stochastic calculus is quadratic variation, presented in Section 3.4. All of this material is needed in order to proceed, except Sections 3.6 and 3.7, which are used only in Chapter 7, *Exotic Options* and Chapter 8, *Early Exercise*.

The core of Volume II is Chapter 4, *Stochastic Calculus*. Here the Itô integral is constructed and Itô's formula (called the Itô-Doeblin formula in this text) is developed. Several consequences of the Itô-Doeblin formula are worked out. One of these is the characterization of Brownian motion in terms of its quadratic variation (Lévy's theorem) and another is the Black-Scholes equation for a European call price (called the Black-Scholes-Merton equation in this text). The only material which the reader may omit is Section 4.7, *Brownian Bridge*. This topic is included because of its importance in Monte Carlo simulation, but it is not used elsewhere in the text.

Chapter 5, *Risk-Neutral Pricing*, states and proves Girsanov's Theorem, which underlies change of measure. This permits a systematic treatment of risk-neutral pricing and the Fundamental Theorems of Asset Pricing (Section 5.4). Section 5.5, *Dividend-Paying Stocks*, is not used elsewhere in the text. Section 5.6, *Forwards and Futures*, appears later in Section 9.4 and in some exercises.

Chapter 6, *Connections with Partial Differential Equations*, develops the connection between stochastic calculus and partial differential equations. This is used frequently in later chapters.

With the exceptions noted above, the material in Chapters 1–6 is fundamental for quantitative finance is essential for reading the later chapters. After Chapter 6, the reader has choices.

Chapter 7, *Exotic Options*, is not used in subsequent chapters, nor is Chapter 8, *Early Exercise*. Chapter 9, *Change of Numéraire*, plays an important role in Section 10.4, *Forward LIBOR model*, but is not otherwise used. Chapter 10, *Term Structure Models*, and Chapter 11, *Introduction to Jump Processes*, are not used elsewhere in the text.

1

The Binomial No-Arbitrage Pricing Model

1.1 One-Period Binomial Model

The *binomial asset-pricing model* provides a powerful tool to understand *arbitrage pricing theory* and probability. In this chapter, we introduce this tool for the first purpose, and we take up the second in Chapter 2. In this section, we consider the simplest binomial model, the one with only one period. This is generalized to the more realistic multiperiod binomial model in the next section.

For the general one-period model of Figure 1.1.1, we call the beginning of the period *time zero* and the end of the period *time one*. At time zero, we have a stock whose price per share we denote by S_0, a positive quantity known at time zero. At time one, the price per share of this stock will be one of two positive values, which we denote $S_1(H)$ and $S_1(T)$, the H and T standing for *head* and *tail*, respectively. Thus, we are imagining that a coin is tossed, and the outcome of the coin toss determines the price at time one. We do not assume this coin is fair (i.e., the probability of head need not be one-half). We assume only that the probability of head, which we call p, is positive, and the probability of tail, which is $q = 1 - p$, is also positive.

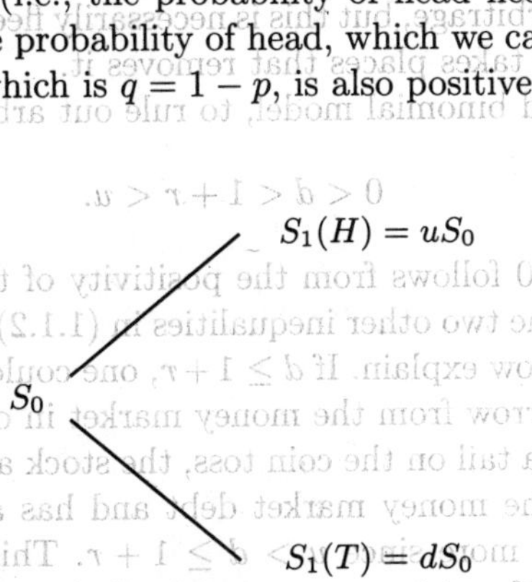

Fig. 1.1.1. General one-period binomial model.

The outcome of the coin toss, and hence the value which the stock price will take at time one, is known at time one but not at time zero. We shall refer to any quantity not known at time zero as *random* because it depends on the random experiment of tossing a coin.

We introduce the two positive numbers

$$u = \frac{S_1(H)}{S_0}, \quad d = \frac{S_1(T)}{S_0}. \tag{1.1.1}$$

We assume that $d < u$; if we instead had $d > u$, we may achieve $d < u$ by relabeling the sides of our coin. If $d = u$, the stock price at time one is not really random and the model is uninteresting. We refer to u as the *up factor* and d as the *down factor*. It is intuitively helpful to think of u as greater than one and to think of d as less than one, and hence the names *up factor* and *down factor*, but the mathematics we develop here does not require that these inequalities hold.

We introduce also an *interest rate* r. One dollar invested in the money market at time zero will yield $1+r$ dollars at time one. Conversely, one dollar borrowed from the money market at time zero will result in a debt of $1+r$ at time one. In particular, the interest rate for borrowing is the same as the interest rate for investing. It is almost always true that $r \geq 0$, and this is the case to keep in mind. However, the mathematics we develop requires only that $r > -1$.

An essential feature of an efficient market is that if a trading strategy can turn nothing into something, then it must also run the risk of loss. Otherwise, there would be an *arbitrage*. More specifically, we define *arbitrage* as a trading strategy that begins with no money, has zero probability of losing money, and has a positive probability of making money. A mathematical model that admits arbitrage cannot be used for analysis. Wealth can be generated from nothing in such a model, and the questions one would want the model to illuminate are provided with paradoxical answers by the model. Real markets sometimes exhibit arbitrage, but this is necessarily fleeting; as soon as someone discovers it, trading takes places that removes it.

In the one-period binomial model, to rule out arbitrage we must assume

$$0 < d < 1 + r < u. \tag{1.1.2}$$

The inequality $d > 0$ follows from the positivity of the stock prices and was already assumed. The two other inequalities in (1.1.2) follow from the absence of arbitrage, as we now explain. If $d \geq 1+r$, one could begin with zero wealth and at time zero borrow from the money market in order to buy stock. Even in the worst case of a tail on the coin toss, the stock at time one will be worth enough to pay off the money market debt and has a positive probability of being worth strictly more since $u > d \geq 1 + r$. This provides an arbitrage. On the other hand, if $u \leq 1 + r$, one could sell the stock short and invest the proceeds in the money market. Even in the best case for the stock, the cost of replacing it at time one will be less than or equal to the value of the money market investment, and since $d < u \le 1+r$, there is a positive probability that the cost of replacing the stock will be strictly less than the value of the money market investment. This again provides an arbitrage.

We have argued in the preceding paragraph that if there is to be no arbitrage in the market with the stock and the money market account, then we must have (1.1.2). The converse of this is also true. If (1.1.2) holds, then there is no arbitrage. See Exercise 1.1.

It is common to have $d = \frac{1}{u}$, and this will be the case in many of our examples. However, for the binomial asset-pricing model to make sense, we only need to assume (1.1.2).

Of course, stock price movements are much more complicated than indicated by the binomial asset-pricing model. We consider this simple model for three reasons. First of all, within this model, the concept of arbitrage pricing and its relation to risk-neutral pricing is clearly illuminated. Secondly, the model is used in practice because, with a sufficient number of periods, it provides a reasonably good, computationally tractable approximation to continuous-time models. Finally, within the binomial asset-pricing model, we can develop the theory of conditional expectations and martingales, which lies at the heart of continuous-time models.

Let us now consider a *European call option*, which confers on its owner the right but not the obligation to buy one share of the stock at time one for the *strike price* K. The interesting case, which we shall assume here, is that $S_1(T) < K < S_1(H)$. If we get a tail on the toss, the option expires worthless. If we get a head on the coin toss, the option can be *exercised* and yields a profit of $S_1(H) - K$. We summarize this situation by saying that the option at time one is worth $(S_1 - K)^+$, where the notation $(\cdots)^+$ indicates that we take the maximum of the expression in parentheses and zero. Here we follow the usual custom in probability of omitting the argument of the random variable S_1. The fundamental question of option pricing is how much the option is worth at time zero before we know whether the coin toss results in head or tail.

The *arbitrage pricing theory* approach to the option-pricing problem is to replicate the option by trading in the stock and money markets. We illustrate this with an example, and then we return to the general one-period binomial model.

Example 1.1.1. For the particular one-period model of Figure 1.1.2, let $S(0) = 4$, $u = 2$, $d = \frac{1}{2}$, and $r = \frac{1}{4}$. Then $S_1(H) = 8$ and $S_1(T) = 2$. Suppose the strike price of the European call option is $K = 5$. Suppose further that we begin with an initial wealth $X_0 = 1.20$ and buy $\Delta_0 = \frac{1}{2}$ shares of stock at time zero. Since stock costs 4 per share at time zero, we must use our initial wealth $X_0 = 1.20$ and borrow an additional 0.80 to do this. This leaves us with a cash position $X_0 - \Delta_0 S_0 = -0.80$ (i.e., a debt of 0.80 to the money market). At time one, our cash position will be $(1+r)(X_0 - \Delta_0 S_0) = -1$

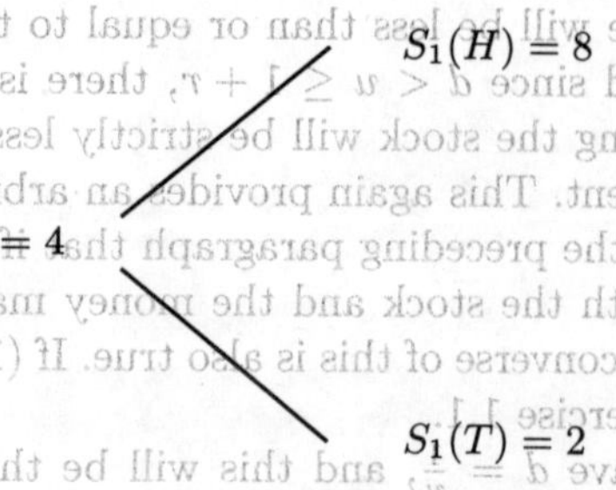

Fig. 1.1.2. Particular one-period binomial model.

(i.e., we will have a debt of 1 to the money market). On the other hand, at time one we will have stock valued at either $\frac{1}{2}S_1(H) = 4$ or $\frac{1}{2}S_1(T) = 1$. In particular, if the coin toss results in a head, the value of our portfolio of stock and money market account at time one will be

$$X_1(H) = \frac{1}{2}S_1(H) + (1+r)(X_0 - \Delta_0 S_0) = 3;$$

if the coin toss results in a tail, the value of our portfolio of stock and money market account at time one will be

$$X_1(T) = \frac{1}{2}S_1(T) + (1+r)(X_0 - \Delta_0 S_0) = 0.$$

In either case, the value of the portfolio agrees with the value of the option at time one, which is either $(S_1(H) - 5)^+ = 3$ or $(S_1(T) - 5)^+ = 0$. We have *replicated* the option by trading in the stock and money markets.

The initial wealth 1.20 needed to set up the replicating portfolio described above is the *no-arbitrage price of the option at time zero*. If one could sell the option for more than this, say, for 1.21, then the seller could invest the excess 0.01 in the money market and use the remaining 1.20 to replicate the option. At time one, the seller would be able to pay off the option, regardless of how the coin tossing turned out, and still have the 0.0125 resulting from the money market investment of the excess 0.01. This is an arbitrage because the seller of the option needs no money initially, and without risk of loss has 0.0125 at time one. On the other hand, if one could buy the option above for less than 1.20, say, for 1.19, then one should buy the option and set up the reverse of the replicating trading strategy described above. In particular, sell short one-half share of stock, which generates income 2. Use 1.19 to buy the option, put 0.80 in the money market, and in a separate money market account put the remaining 0.01. At time one, if there is a head, one needs 4 to replace the half-share of stock. The option bought at time zero is worth 3, and the 0.80 invested in the money market at time zero has grown to 1. At time one, if there is a tail, one needs 1 to replace the half-share of stock.

The option is worthless, but the 0.80 invested in the money market at time zero has grown to 1. In either case, the buyer of the option has a net zero position at time one, plus the separate money market account in which 0.01 was invested at time zero. Again, there is an arbitrage.

We have shown that in the market with the stock, the money market, and the option, there is an arbitrage unless the time-zero price of the option is 1.20. If the time-zero price of the option is 1.20, then there is no arbitrage (see Exercise 1.2). □

The argument in the example above depends on several assumptions. The principal ones are:

- shares of stock can be subdivided for sale or purchase,
- the interest rate for investing is the same as the interest rate for borrowing,
- the purchase price of stock is the same as the selling price (i.e., there is zero ***bid–ask spread***),
- at any time, the stock can take only two possible values in the next period.

All these assumptions except the last also underlie the Black-Scholes-Merton option-pricing formula. The first of these assumptions is essentially satisfied in practice because option pricing and hedging (replication) typically involve lots of options. If we had considered 100 options rather than one option in Example 1.1.1, we would have hedged the short position by buying $\Delta_0 = 50$ shares of stock rather than $\Delta_0 = \frac{1}{2}$ of a share. The second assumption is close to being true for large institutions. The third assumption is not satisfied in practice. Sometimes the bid–ask spread can be ignored because not too much trading is taking place. In other situations, this departure of the model from reality becomes a serious issue. In the Black-Scholes-Merton model, the fourth assumption is replaced by the assumption that the stock price is a geometric Brownian motion. Empirical studies of stock price returns have consistently shown this not to be the case. Once again, the departure of the model from reality can be significant in some situations, but in other situations the model works remarkably well. We shall develop a modeling framework that extends far beyond the geometric Brownian motion assumption, a framework that includes many of the more sophisticated models that are not tied to this assumption.

In the general one-period model, we define a *derivative security* to be a security that pays some amount $V_1(H)$ at time one if the coin toss results in head and pays a possibly different amount $V_1(T)$ at time one if the coin toss results in tail. A European call option is a particular kind of derivative security. Another is the *European put option*, which pays off $(K - S_1)^+$ at time one, where K is a constant. A third is a *forward contract*, whose value at time one is $S_1 - K$.

To determine the price V_0 at time zero for a derivative security, we replicate it as in Example 1.1.1. Suppose we begin with wealth X_0 and buy Δ_0 shares of stock at time zero, leaving us with a cash position $X_0 - \Delta_0 S_0$. The value of our portfolio of stock and money market account at time one is

$$X_1 = \Delta_0 S_1 + (1+r)(X_0 - \Delta_0 S_0) = (1+r)X_0 + \Delta_0(S_1 - (1+r)S_0).$$

We want to choose X_0 and Δ_0 so that $X_1(H) = V_1(H)$ and $X_1(T) = V_1(T)$. (Note here that $V_1(H)$ and $V_1(T)$ are given quantities, the amounts the derivative security will pay off depending on the outcome of the coin tosses. At time zero, we know what the two possibilities $V_1(H)$ and $V_1(T)$ are; we do not know which of these two possibilities will be realized.) Replication of the derivative security thus requires that

$$X_0 + \Delta_0\left(\frac{1}{1+r}S_1(H) - S_0\right) = \frac{1}{1+r}V_1(H), \tag{1.1.3}$$

$$X_0 + \Delta_0\left(\frac{1}{1+r}S_1(T) - S_0\right) = \frac{1}{1+r}V_1(T). \tag{1.1.4}$$

One way to solve these two equations in two unknowns is to multiply the first by a number $\tilde{p}$ and the second by $\tilde{q} = 1 - \tilde{p}$ and then add them to get

$$X_0 + \Delta_0\left(\frac{1}{1+r}\left[\tilde{p}S_1(H) + \tilde{q}S_1(T)\right] - S_0\right) = \frac{1}{1+r}\left[\tilde{p}V_1(H) + \tilde{q}V_1(T)\right]. \tag{1.1.5}$$

If we choose $\tilde{p}$ so that

$$S_0 = \frac{1}{1+r}\left[\tilde{p}S_1(H) + \tilde{q}S_1(T)\right], \tag{1.1.6}$$

then the term multiplying Δ_0 in (1.1.5) is zero, and we have the simple formula for X_0

$$X_0 = \frac{1}{1+r}\left[\tilde{p}V_1(H) + \tilde{q}V_1(T)\right]. \tag{1.1.7}$$

We can solve for $\tilde{p}$ directly from (1.1.6) in the form

$$S_0 = \frac{1}{1+r}\left[\tilde{p}uS_0 + (1-\tilde{p})dS_0\right] = \frac{S_0}{1+r}\left[(u-d)\tilde{p} + d\right].$$

This leads to the formulas

$$\tilde{p} = \frac{1+r-d}{u-d}, \quad \tilde{q} = \frac{u-1-r}{u-d}. \tag{1.1.8}$$

We can solve for Δ_0 by simply subtracting (1.1.4) from (1.1.3) to get the *delta-hedging formula*

$$\Delta_0 = \frac{V_1(H) - V_1(T)}{S_1(H) - S_1(T)}. \tag{1.1.9}$$

In conclusion, if an agent begins with wealth X_0 given by (1.1.7) and at time zero buys Δ_0 shares of stock, given by (1.1.9), then at time one, if the coin toss results in head, the agent will have a portfolio worth $V_1(H)$, and if the coin toss results in tail, the portfolio will be worth $V_1(T)$. The agent has *hedged a short position in the derivative security.* The derivative security that pays V_1 at time one should be priced at

$$V_0 = \frac{1}{1+r}\left[\tilde{p}V_1(H) + \tilde{q}V_1(T)\right] \tag{1.1.10}$$

at time zero. This price permits the seller to hedge the short position in the claim. This price does not introduce an arbitrage when the derivative security is added to the market comprising the stock and money market account; any other time-zero price would introduce an arbitrage.

Although we have determined the no-arbitrage price of a derivative security by setting up a hedge for a short position in the security, one could just as well consider the hedge for a long position. An agent with a long position owns an asset having a certain value, and the agent may wish to set up a hedge to protect against loss of that value. This is how practitioners think about hedging. The number of shares of the underlying stock held by a long position hedge is the negative of the number determined by (1.1.9). Exercises 1.6 and 1.7 consider this in more detail.

The numbers $\tilde{p}$ and $\tilde{q}$ given by (1.1.8) are both positive because of the no-arbitrage condition (1.1.2), and they sum to one. For this reason, we can regard them as probabilities of head and tail, respectively. They are not the actual probabilities, which we call p and q, but rather the so-called *risk-neutral probabilities.* Under the actual probabilities, the average rate of growth of the stock is typically strictly greater than the rate of growth of an investment in the money market; otherwise, no one would want to incur the risk associated with investing in the stock. Thus, p and $q = 1 - p$ should satisfy

$$S_0 < \frac{1}{1+r}\left[pS_1(H) + qS_1(T)\right],$$

whereas $\tilde{p}$ and $\tilde{q}$ satisfy (1.1.6). If the average rate of growth of the stock were exactly the same as the rate of growth of the money market investment, then investors must be neutral about risk—they do not require compensation for assuming it, nor are they willing to pay extra for it. This is simply not the case, and hence $\tilde{p}$ and $\tilde{q}$ cannot be the actual probabilities. They are only numbers that assist us in the solution of the two equations (1.1.3) and (1.1.4) in the two unknowns X_0 and Δ_0. They assist us by making the term multiplying the unknown Δ_0 in (1.1.5) drop out. In fact, because they are chosen to make the mean rate of growth of the stock appear to equal the rate of growth of the money market account, they make the mean rate of growth of any portfolio of stock and money market account appear to equal the rate of growth of the money market asset. If we want to construct a portfolio whose value at time one is V_1, then its value at time zero must be given by (1.1.7), so that its mean rate of growth under the risk-neutral probabilities is the rate of growth of the money market investment.

The concluding equation (1.1.10) for the time-zero price V_0 of the derivative security V_1 is called the *risk-neutral pricing formula* for the one-period binomial model. One should not be concerned that the actual probabilities do not appear in this equation. We have constructed a hedge for a short position in the derivative security, and this hedge works regardless of whether the stock goes up or down. The probabilities of the up and down moves are irrelevant. What matters is the size of the two possible moves (the values of u and d). In the binomial model, the prices of derivative securities depend on the set of possible stock price paths but not on how probable these paths are. As we shall see in Chapters 4 and 5 of Volume II, the analogous fact for continuous-time models is that prices of derivative securities depend on the volatility of stock prices but not on their mean rates of growth.

1.2 Multiperiod Binomial Model

We now extend the ideas in Section 1.1 to multiple periods. We toss a coin repeatedly, and whenever we get a head the stock price moves "up" by the factor u, whereas whenever we get a tail, the stock price moves "down" by the factor d. In addition to this stock, there is a money market asset with a constant interest rate r. The only assumption we make on these parameters is the no-arbitrage condition (1.1.2).

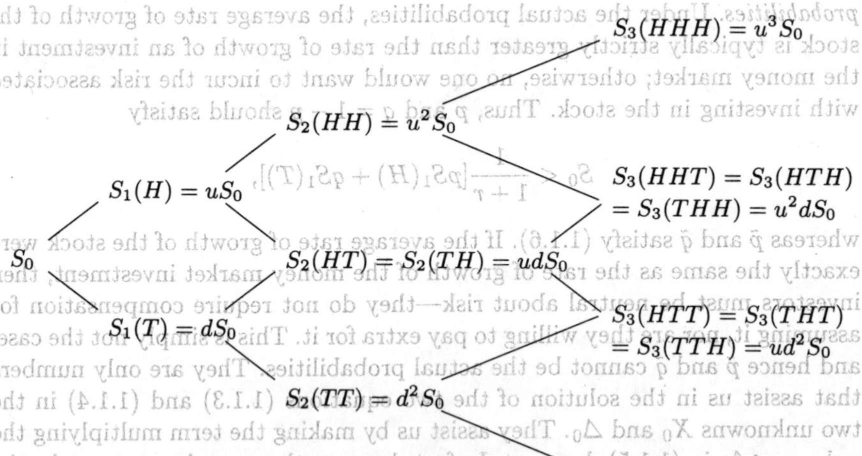

Fig. 1.2.1. General three-period model.

We denote the initial stock price by S_0, which is positive. We denote the price at time one by $S_1(H) = uS_0$ if the first toss results in head and by $S_1(T) = dS_0$ if the first toss results in tail. After the second toss, the price will be one of:

$$S_2(HH) = uS_1(H) = u^2S_0, \; S_2(HT) = dS_1(H) = duS_0,$$
$$S_2(TH) = uS_1(T) = udS_0, \; S_2(TT) = dS_1(T) = d^2S_0.$$

After three tosses, there are eight possible coin sequences, although not all of them result in different stock prices at time 3. See Figure 1.2.1.

Example 1.2.1. Consider the particular three-period model with $S_0 = 4$, $u = 2$, and $d = \frac{1}{2}$. We have the binomial "tree" of possible stock prices shown in Figure 1.2.2. □

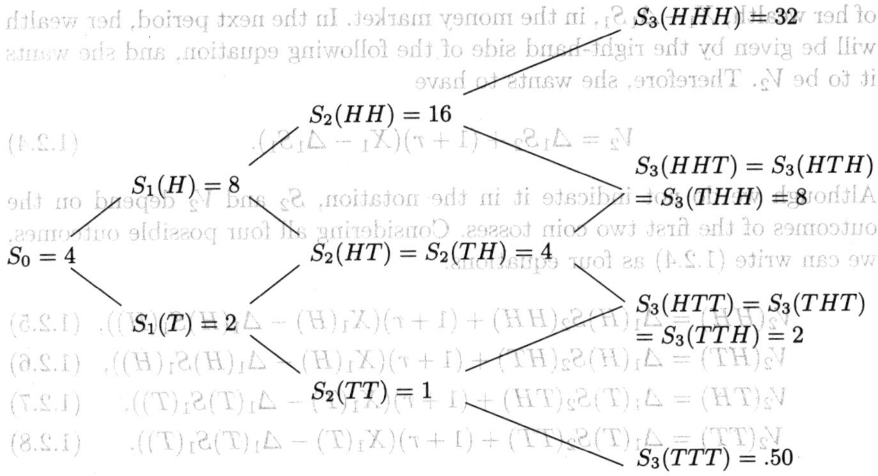

Fig. 1.2.2. A particular three-period model.

Let us return to the general three-period binomial model of Figure 1.2.1 and consider a European call that confers the right to buy one share of stock for K dollars at time two. After the discussion of this option, we extend the analysis to an arbitrary European derivative security that expires at time $N \geq 2$.

At expiration, the payoff of a call option with strike price K and expiration time two is $V_2 = (S_2 - K)^+$, where V_2 and S_2 depend on the first and second coin tosses. We want to determine the no-arbitrage price for this option at time zero. Suppose an agent sells the option at time zero for V_0 dollars, where V_0 is still to be determined. She then buys Δ_0 shares of stock, investing $V_0 - \Delta_0 S_0$ dollars in the money market to finance this. (The quantity $V_0 - \Delta_0 S_0$ will turn out to be negative, so the agent is actually borrowing $\Delta_0 S_0 - V_0$ dollars from the money market.) At time one, the agent has a portfolio (excluding the short position in the option) valued at

$$X_1 = \Delta_0 S_1 + (1 + r)(V_0 - \Delta_0 S_0). \tag{1.2.1}$$

Although we do not indicate it in the notation, S_1 and therefore X_1 depend on the outcome of the first coin toss. Thus, there are really two equations implicit in (1.2.1):

$$X_1(H) = \Delta_0 S_1(H) + (1+r)(V_0 - \Delta_0 S_0), \tag{1.2.2}$$
$$X_1(T) = \Delta_0 S_1(T) + (1+r)(V_0 - \Delta_0 S_0). \tag{1.2.3}$$

After the first coin toss, the agent has a portfolio valued at X_1 dollars and can readjust her hedge. Suppose she decides now to hold Δ_1 shares of stock, where Δ_1 is allowed to depend on the first coin toss because the agent knows the result of this toss at time one when she chooses Δ_1. She invests the remainder of her wealth, $X_1 - \Delta_1 S_1$, in the money market. In the next period, her wealth will be given by the right-hand side of the following equation, and she wants it to be V_2. Therefore, she wants to have

$$V_2 = \Delta_1 S_2 + (1+r)(X_1 - \Delta_1 S_1). \tag{1.2.4}$$

Although we do not indicate it in the notation, S_2 and V_2 depend on the outcomes of the first two coin tosses. Considering all four possible outcomes, we can write (1.2.4) as four equations:

$$V_2(HH) = \Delta_1(H)S_2(HH) + (1+r)(X_1(H) - \Delta_1(H)S_1(H)), \tag{1.2.5}$$
$$V_2(HT) = \Delta_1(H)S_2(HT) + (1+r)(X_1(H) - \Delta_1(H)S_1(H)), \tag{1.2.6}$$
$$V_2(TH) = \Delta_1(T)S_2(TH) + (1+r)(X_1(T) - \Delta_1(T)S_1(T)), \tag{1.2.7}$$
$$V_2(TT) = \Delta_1(T)S_2(TT) + (1+r)(X_1(T) - \Delta_1(T)S_1(T)). \tag{1.2.8}$$

We now have six equations, the two represented by (1.2.1) and the four represented by (1.2.4), in the six unknowns V_0, Δ_0, $\Delta_1(H)$, $\Delta_1(T)$, $X_1(H)$, and $X_1(T)$.

To solve these equations, and thereby determine the no-arbitrage price V_0 at time zero of the option and the replicating portfolio Δ_0, $\Delta_1(H)$, and $\Delta_1(T)$, we begin with the last two equations, (1.2.7) and (1.2.8). Subtracting (1.2.8) from (1.2.7) and solving for $\Delta_1(T)$, we obtain the *delta-hedging formula*

$$\Delta_1(T) = \frac{V_2(TH) - V_2(TT)}{S_2(TH) - S_2(TT)}, \tag{1.2.9}$$

and substituting this into either (1.2.7) or (1.2.8), we can solve for

$$X_1(T) = \frac{1}{1+r}[\tilde{p}V_2(TH) + \tilde{q}V_2(TT)], \tag{1.2.10}$$

where $\tilde{p}$ and $\tilde{q}$ are the risk-neutral probabilities given by (1.1.8). We can also obtain (1.2.10) by multiplying (1.2.7) by $\tilde{p}$ and (1.2.8) by $\tilde{q}$ and adding them together. Since

$$\tilde{p}S_2(TH) + \tilde{q}S_2(TT) = (1+r)S_1(T),$$

this causes all the terms involving $\Delta_1(T)$ to drop out. Equation (1.2.10) gives the value the replicating portfolio should have at time one if the stock goes down between times zero and one. We define this quantity to be the *price of the option at time one if the first coin toss results in tail*, and we denote it by $V_1(T)$. We have just shown that

$$V_1(T) = \frac{1}{1+r}[\tilde{p}V_2(TH) + \tilde{q}V_2(TT)], \tag{1.2.11}$$

which is another instance of the *risk-neutral pricing formula*. This formula is analogous to formula (1.1.10) but postponed by one period. The first two equations, (1.2.5) and (1.2.6), lead in a similar way to the formulas

$$\Delta_1(H) = \frac{V_2(HH) - V_2(HT)}{S_2(HH) - S_2(HT)} \tag{1.2.12}$$

and $X_1(H) = V_1(H)$, where $V_1(H)$ is the *price of the option at time one if the first toss results in head*, defined by

$$V_1(H) = \frac{1}{1+r}[\tilde{p}V_2(HH) + \tilde{q}V_2(HT)]. \tag{1.2.13}$$

This is again analogous to formula (1.1.10), postponed by one period. Finally, we plug the values $X_1(H) = V_1(H)$ and $X_1(T) = V_1(T)$ into the two equations implicit in (1.2.1). The solution of these equations for Δ_0 and V_0 is the same as the solution of (1.1.3) and (1.1.4) and results again in (1.1.9) and (1.1.10).

To recap, we have three *stochastic processes*, (Δ_0, Δ_1), (X_0, X_1, X_2), and (V_0, V_1, V_2). By *stochastic process*, we mean a sequence of random variables indexed by time. These quantities are random because they depend on the coin tosses; indeed, the subscript on each variable indicates the number of coin tosses on which it depends. If we begin with any initial wealth X_0 and specify values for Δ_0, $\Delta_1(H)$, and $\Delta_1(T)$, then we can compute the value of the portfolio that holds the number of shares of stock indicated by these specifications and finances this by borrowing or investing in the money market as necessary. Indeed, the value of this portfolio is defined recursively, beginning with X_0, via the *wealth equation*

$$X_{n+1} = \Delta_n S_{n+1} + (1+r)(X_n - \Delta_n S_n). \tag{1.2.14}$$

One might regard this as a contingent equation; it defines *random variables*, and actual values of these random variables are not resolved until the outcomes of the coin tossing are revealed. Nonetheless, already at time zero this equation permits us to compute what the value of the portfolio will be at every subsequent time under every coin-toss scenario.

For a derivative security expiring at time two, the random variable V_2 is contractually specified in a way that is contingent upon the coin tossing (e.g., if the coin tossing results in $\omega_1\omega_2$, so the stock price at time two is $S_2(\omega_1\omega_2)$, then for the European call we have $V_2(\omega_1\omega_2) = (S_2(\omega_1\omega_2) - K)^+$). We want to determine a value of X_0 and values for Δ_0, $\Delta_1(H)$, and $\Delta_1(T)$ so that X_2 given by applying (1.2.14) recursively satisfies $X_2(\omega_1\omega_2) = V_2(\omega_1\omega_2)$, regardless of the values of ω_1 and ω_2. The formulas above tell us how to do this. We call V_0 the value of X_0 that allows us to accomplish this, and we define $V_1(H)$ and $V_1(T)$ to be the values of $X_1(H)$ and $X_1(T)$ given by (1.2.14) when X_0 and Δ_0 are chosen by the prescriptions above. In general, we use the symbols Δ_n and X_n to represent the number of shares of stock held by the portfolio and the corresponding portfolio values, respectively, regardless of how the initial wealth X_0 and the Δ_n are chosen. When X_0 and the Δ_n are chosen to replicate a derivative security, we use the symbol V_n in place of X_n and call this the *(no-arbitrage) price of the derivative security at time n.*

The pattern that emerged with the European call expiring at time two persists, regardless of the number of periods and the definition of the final payoff of the derivative security. (At this point, however, we are considering only payoffs that come at a specified time; there is no possibility of early exercise.)

Theorem 1.2.2 (Replication in the multiperiod binomial model). *Consider an N-period binomial asset-pricing model, with $0 < d < 1 + r < u$, and with*

$$\tilde{p} = \frac{1+r-d}{u-d}, \quad \tilde{q} = \frac{u-1-r}{u-d}. \tag{1.2.15}$$

Let V_N be a random variable (a derivative security paying off at time N) depending on the first N coin tosses $\omega_1\omega_2\ldots\omega_N$. Define recursively backward in time the sequence of random variables $V_{N-1}, V_{N-2}, \ldots, V_0$ by

$$V_n(\omega_1\omega_2\ldots\omega_n) = \frac{1}{1+r}[\tilde{p}V_{n+1}(\omega_1\omega_2\ldots\omega_nH) + \tilde{q}V_{n+1}(\omega_1\omega_2\ldots\omega_nT)], \tag{1.2.16}$$

so that each V_n depends on the first n coin tosses $\omega_1\omega_2\ldots\omega_n$, where n ranges between $N-1$ and 0. Next define

$$\Delta_n(\omega_1\ldots\omega_n) = \frac{V_{n+1}(\omega_1\ldots\omega_nH) - V_{n+1}(\omega_1\ldots\omega_nT)}{S_{n+1}(\omega_1\ldots\omega_nH) - S_{n+1}(\omega_1\ldots\omega_nT)}, \tag{1.2.17}$$

where again n ranges between 0 and $N-1$. If we set $X_0 = V_0$ and define recursively forward in time the portfolio values $X_1, X_2, \ldots, X_N$ by (1.2.14), then we will have

$$X_N(\omega_1\omega_2\ldots\omega_N) = V_N(\omega_1\omega_2\ldots\omega_N) \text{ for all } \omega_1\omega_2\ldots\omega_N. \tag{1.2.18}$$

Definition 1.2.3. *For $n = 1, 2, \ldots, N$, the* price of the derivative security *at time n if the outcomes of the first n tosses are $\omega_1\ldots\omega_n$ is defined to be the random variable $V_n(\omega_1\ldots\omega_n)$ of Theorem 1.2.2. The* price of the derivative security at time zero *is defined to be V_0.*

PROOF OF THEOREM 1.2.2: We prove by forward induction on n that

$$X_n(\omega_1\omega_2\dots\omega_n) = V_n(\omega_1\omega_2\dots\omega_n) \text{ for all } \omega_1\omega_2\dots\omega_n, \tag{1.2.19}$$

where n ranges between 0 and N. The case of $n = 0$ is given by the definition of X_0 as V_0. The case of $n = N$ is what we want to show.

For the induction step, we assume that (1.2.19) holds for some value of n less than N and show that it holds for $n+1$. We thus let $\omega_1\omega_2\dots\omega_n\omega_{n+1}$ be fixed but arbitrary and assume as the induction hypothesis that (1.2.19) holds for the particular $\omega_1\omega_2\dots\omega_n$ we have fixed. We don't know whether $\omega_{n+1} = H$ or $\omega_{n+1} = T$, so we consider both cases. We first use (1.2.14) to compute $X_{n+1}(\omega_1\omega_2\dots\omega_n H)$, to wit

$$\begin{aligned} &X_{n+1}(\omega_1\omega_2\dots\omega_n H)\\ &= \Delta_n(\omega_1\omega_2\dots\omega_n)uS_n(\omega_1\omega_2\dots\omega_n)\\ &\quad +(1+r)\Big(X_n(\omega_1\omega_2\dots\omega_n) - \Delta_n(\omega_1\omega_2\dots\omega_n)S_n(\omega_1\omega_2\dots\omega_n)\Big). \end{aligned}$$

To simplify the notation, we suppress $\omega_1\omega_2\dots\omega_n$ and write this equation simply as

$$X_{n+1}(H) = \Delta_n u S_n + (1+r)(X_n - \Delta_n S_n). \tag{1.2.20}$$

With $\omega_1\omega_2\dots\omega_n$ similarly suppressed, we have from (1.2.17) that

$$\Delta_n = \frac{V_{n+1}(H) - V_{n+1}(T)}{S_{n+1}(H) - S_{n+1}(T)} = \frac{V_{n+1}(H) - V_{n+1}(T)}{(u-d)S_n}.$$

Substituting this into (1.2.20) and using the induction hypothesis (1.2.19) and the definition (1.2.16) of V_n, we see that

$$\begin{aligned} X_{n+1}(H) &= (1+r)X_n + \Delta_n S_n(u-(1+r))\\ &= (1+r)V_n + \frac{(V_{n+1}(H) - V_{n+1}(T))(u-(1+r))}{u-d}\\ &= (1+r)V_n + \tilde{q}V_{n+1}(H) - \tilde{q}V_{n+1}(T)\\ &= \tilde{p}V_{n+1}(H) + \tilde{q}V_{n+1}(T) + \tilde{q}V_{n+1}(H) - \tilde{q}V_{n+1}(T)\\ &= V_{n+1}(H). \end{aligned}$$

Reinstating the suppressed coin tosses $\omega_1\omega_2\dots\omega_n$, we may write this as

$$X_{n+1}(\omega_1\omega_2\dots\omega_n H) = V_{n+1}(\omega_1\omega_2\dots\omega_n H).$$

A similar argument (see Exercise 1.4) shows that

$$X_{n+1}(\omega_1\omega_2\dots\omega_n T) = V_{n+1}(\omega_1\omega_2\dots\omega_n T).$$

Consequently, regardless of whether $\omega_{n+1} = H$ or $\omega_{n+1} = T$, we have

$$X_{n+1}(\omega_1\omega_2\dots\omega_n\omega_{n+1}) = V_{n+1}(\omega_1\omega_2\dots\omega_n\omega_{n+1}).$$

Since $\omega_1\omega_2\ldots\omega_n\omega_{n+1}$ is arbitrary, the induction step is complete. □

The multiperiod binomial model of this section is said to be *complete* because every derivative security can be replicated by trading in the underlying stock and the money market. In a complete market, every derivative security has a unique price that precludes arbitrage, and this is the price of Definition 1.2.3.

Theorem 1.2.2 applies to so-called *path-dependent* options as well as to derivative securities whose payoff depends only on the final stock price. We illustrate this point with the following example.

Example 1.2.4. Suppose as in Figure 1.2.2 that $S_0 = 4$, $u = 2$, and $d = \frac{1}{2}$. Assume the interest rate is $r = \frac{1}{4}$. Then $\tilde{p} = \tilde{q} = \frac{1}{2}$. Consider a *lookback option* that pays off

$$V_3 = \max_{0\leq n\leq 3} S_n - S_3$$

at time three. Then

$$\begin{aligned}
V_3(HHH) &= S_3(HHH) - S_3(HHH) = 32 - 32 = 0,\\
V_3(HHT) &= S_2(HH) - S_3(HHT) = 16 - 8 = 8,\\
V_3(HTH) &= S_1(H) - S_3(HTH) = 8 - 8 = 0,\\
V_3(HTT) &= S_1(H) - S_3(HTT) = 8 - 2 = 6,\\
V_3(THH) &= S_3(THH) - S_3(THH) = 8 - 8 = 0,\\
V_3(THT) &= S_2(TH) - S_3(THT) = 4 - 2 = 2,\\
V_3(TTH) &= S_0 - S_3(TTH) = 4 - 2 = 2,\\
V_3(TTT) &= S_0 - S_3(TTT) = 4 - 0.50 = 3.50.
\end{aligned}$$

We compute the price of the option at other times using the backward recursion (1.2.16). This gives

$$V_2(HH) = \frac{4}{5}\left[\frac{1}{2}V_3(HHH) + \frac{1}{2}V_3(HHT)\right] = 3.20,$$

$$V_2(HT) = \frac{4}{5}\left[\frac{1}{2}V_3(HTH) + \frac{1}{2}V_3(HTT)\right] = 2.40,$$

$$V_2(TH) = \frac{4}{5}\left[\frac{1}{2}V_3(THH) + \frac{1}{2}V_3(THT)\right] = 0.80,$$

$$V_2(TT) = \frac{4}{5}\left[\frac{1}{2}V_3(TTH) + \frac{1}{2}V_3(TTT)\right] = 2.20,$$

and then

$$V_1(H) = \frac{4}{5}\left[\frac{1}{2}V_2(HH) + \frac{1}{2}V_2(HT)\right] = 2.24,$$

$$V_1(T) = \frac{4}{5}\left[\frac{1}{2}V_2(TH) + \frac{1}{2}V_2(TT)\right] = 1.20,$$

and finally

$$V_0 = \frac{4}{5}\left[\frac{1}{2}V_1(H) + \frac{1}{2}V_1(T)\right] = 1.376.$$

If an agent sells the lookback option at time zero for 1.376, she can hedge her short position in the option by buying

$$\Delta_0 = \frac{V_1(H) - V_1(T)}{S_1(H) - S_1(T)} = \frac{2.24 - 1.20}{8 - 2} = 0.1733$$

shares of stock. This costs 0.6933 dollars, which leaves her with 1.376 − 0.6933 = 0.6827 to invest in the money market at 25% interest. At time one, she will have 0.8533 in the money market. If the stock goes up in price to 8, then at time one her stock is worth 1.3867, and so her total portfolio value is 2.24, which is $V_1(H)$. If the stock goes down in price to 2, then at time one her stock is worth 0.3467 and so her total portfolio value is 1.20, which is $V_1(T)$. Continuing this process, the agent can be sure to have a portfolio worth V_3 at time three, no matter how the coin tossing turns out. □

1.3 Computational Considerations

The amount of computation required by a naive implementation of the derivative security pricing algorithm given in Theorem 1.2.2 grows exponentially with the number of periods. The binomial models used in practice often have 100 or more periods, and there are $2^{100} \approx 10^{30}$ possible outcomes for a sequence of 100 coin tosses. An algorithm that begins by tabulating 2^{100} values for V_{100} is not computationally practical.

Fortunately, the algorithm given in Theorem 1.2.2 can usually be organized in a computationally efficient manner. We illustrate this with two examples.

Example 1.3.1. In the model with $S_0 = 4$, $u = 2$, $d = \frac{1}{2}$ and $r = \frac{1}{4}$, consider the problem of pricing a European put with strike price $K = 5$, expiring at time three. The risk-neutral probabilities are $\tilde{p} = \frac{1}{2}$, $\tilde{q} = \frac{1}{2}$. The stock process is shown in Figure 1.2.2. The payoff of the option, given by $V_3 = (5 - S_3)^+$, can be tabulated as

$$V_3(HHH) = 0, \quad V_3(HHT) = V_3(HTH) = V_3(THH) = 0$$
$$V_3(HTT) = V_3(THT) = V_3(TTH) = 3, \quad V_3(TTT) = 4.50.$$

There are $2^3 = 8$ entries in this table, but an obvious simplification is possible. Let us denote by $v_3(s)$ the payoff of the option at time three when the stock price at time three is s. Whereas V_3 has the sequence of three coin tosses as its argument, the argument of v_3 is a stock price. At time three there are only four possible stock prices, and we can tabulate the relevant values of v_3 as

$$v_3(32) = 0, \; v_3(8) = 0, \; v_3(2) = 3, \; v_3(.50) = 4.50.$$

If the put expired after 100 periods, the argument of V_{100} would range over the 2^{100} possible outcomes of the coin tosses whereas the argument of v_{100} would range over the 101 possible stock prices at time 100. This is a tremendous reduction in computational complexity.

According to Theorem 1.2.2, we compute V_2 by the formula

$$V_2(\omega_1\omega_2) = \frac{2}{5}\Big[V_3(\omega_1\omega_2H) + V_3(\omega_1\omega_2T)\Big]. \tag{1.3.1}$$

Equation (1.3.1) represents four equations, one for each possible choice of $\omega_1\omega_2$. We let $v_2(s)$ denote the price of the put at time two if the stock price at time two is s. In terms of this function, (1.3.1) takes the form

$$v_2(s) = \frac{2}{5}\Big[v_3(2s) + v_3(\tfrac{1}{2}s)\Big],$$

and this represents only three equations, one for each possible value of the stock price at time two. Indeed, we may compute

$$v_2(16) = \frac{2}{5}\Big[v_3(32) + v_3(8)\Big] = 0,$$

$$v_2(4) = \frac{2}{5}\Big[v_3(8) + v_3(2)\Big] = 1.20,$$

$$v_2(1) = \frac{2}{5}\Big[v_3(2) + v_3(.50)\Big] = 3.$$

Similarly,

$$v_1(8) = \frac{2}{5}\Big[v_2(16) + v_2(4)\Big] = 0.48,$$

$$v_1(2) = \frac{2}{5}\Big[v_2(4) + v_2(1)\Big] = 1.68,$$

where $v_1(s)$ denotes the price of the put at time one if the stock price at time one is s. The price of the put at time zero is

$$v_0(4) = \frac{2}{5}\Big[v_1(8) + v_1(2)\Big] = 0.864.$$

At each time $n = 0, 1, 2$, if the stock price is s, the number of shares of stock that should be held by the replicating portfolio is

$$\delta_n(s) = \frac{v_{n+1}(2s) - v_{n+1}(\frac{1}{2}s)}{2s - \frac{1}{2}s}.$$

This is the analogue of formula (1.2.17). □

In Example 1.3.1, the price of the option at any time n was a function of the stock price S_n at that time and did not otherwise depend on the coin tosses. This permitted the introduction of the functions v_n related to the random variables V_n by the formula $V_n = v_n(S_n)$. A similar reduction is often possible when the price of the option does depend on the stock price path rather than just the current stock price. We illustrate this with a second example.

Example 1.3.2. Consider the lookback option of Example 1.2.4. At each time n, the price of the option can be written as a function of the stock price S_n and the maximum stock price $M_n = \max_{0 \le k \le n} S_k$ to date. At time three, there are six possible pairs of values for (S_3, M_3), namely

$$(32, 32),\ (8, 16),\ (8, 8),\ (2, 8),\ (2, 4),\ (.50, 4).$$

We define $v_3(s, m)$ to be the payoff of the option at time three if $S_3 = s$ and $M_3 = m$. We have

$$v_3(32, 32) = 0,\ v_3(8, 16) = 8,\ v_3(8, 8) = 0,$$
$$v_3(2, 8) = 6,\ v_3(2, 4) = 2,\ v_3(.50, 4) = 3.50.$$

In general, let $v_n(s, m)$ denote the value of the option at time n if $S_n = s$ and $M_n = m$. The algorithm of Theorem 1.2.2 can be rewritten in terms of the functions v_n as

$$v_n(s, m) = \frac{2}{5}\Big[v_{n+1}(2s, m \vee (2s)) + v_{n+1}(\tfrac{1}{2}s, m)\Big],$$

where $m \vee (2s)$ denotes the maximum of m and $2s$. Using this algorithm, we compute

$$\begin{aligned}
v_2(16, 16) &= \frac{2}{5}\Big[v_3(32, 32) + v_3(8, 16)\Big] = 3.20,\\
v_2(4, 8) &= \frac{2}{5}\Big[v_3(8, 8) + v_3(2, 8)\Big] = 2.40,\\
v_2(4, 4) &= \frac{2}{5}\Big[v_3(8, 8) + v_3(2, 4)\Big] = 0.80,\\
v_2(1, 4) &= \frac{2}{5}\Big[v_3(2, 4) + v_3(.50, 4)\Big] = 2.20,
\end{aligned}$$

then compute

$$\begin{aligned}
v_1(8, 8) &= \frac{2}{5}\Big[v_2(16, 16) + v_2(4, 8)\Big] = 2.24,\\
v_1(2, 4) &= \frac{2}{5}\Big[v_1(4, 4) + v_1(1, 4)\Big] = 1.20,
\end{aligned}$$

and finally obtain the time-zero price

$$v_0(4, 4) = \frac{2}{5}\Big[v_1(8, 8) + v_1(2, 4)\Big] = 1.376.$$

At each time $n = 0, 1, 2$, if the stock price is s and the maximum stock price to date is m, the number of shares of stock that should be held by the replicating portfolio is

$$\delta_n(s, m) = \frac{v_{n+1}(2s, m \vee (2s)) - v_{n+1}(\frac{1}{2}s, m)}{2s - \frac{1}{2}s}.$$

This is the analogue of formula (1.2.17). □

1.4 Summary

This chapter considers a multiperiod binomial model. At each period in this model, we toss a coin whose outcome determines whether the stock price changes by a factor of u or a factor of d, where $0 < d < u$. In addition to the stock, there is a money market account with per-period rate of interest r. This is the rate of interest applied to both investing and borrowing.

Arbitrage is a trading strategy that begins with zero capital and trades in the stock and money markets in order to make money with positive probability without any possibility of losing money. The multiperiod binomial model admits no arbitrage if and only if

$$0 < d < 1 + r < u. \tag{1.1.2}$$

We shall always impose this condition.

A derivative security pays off at some expiration time N contingent upon the coin tosses in the first N periods. The *arbitrage pricing theory* method of assigning a price to a derivative security prior to expiration can be understood in two ways. First, one can ask how to assign a price so that one cannot form an arbitrage by trading in the derivative security, the underlying stock, and the money market. This no-arbitrage condition uniquely determines the price at all times of the derivative security. Secondly, at any time n prior to the expiration time N, one can imagine selling the derivative security for a price and using the income from this sale to form a portfolio, dynamically trading the stock and money market asset from time n until the expiration time N. This portfolio hedges the short position in the derivative security if its value at time N agrees with the payoff of the derivative security, regardless of the outcome of the coin tossing between times n and N. The amount for which the derivative security must be sold at time n in order to construct this hedge of the short position is the same no-arbitrage price obtained by the first pricing method.

The no-arbitrage price of the derivative security that pays V_N at time N can be computed recursively, backward in time, by the formula

$$V_n(\omega_1\omega_2\dots\omega_n) = \frac{1}{1+r}[\tilde{p}V_{n+1}(\omega_1\omega_2\dots\omega_n H) + \tilde{q}V_{n+1}(\omega_1\omega_2\dots\omega_n T)]. \tag{1.2.16}$$

The number of shares of the stock that should be held by a portfolio hedging a short position in the derivative security is given by

$$\Delta_n(\omega_1 \dots \omega_n) = \frac{V_{n+1}(\omega_1 \dots \omega_n H) - V_{n+1}(\omega_1 \dots \omega_n T)}{S_{n+1}(\omega_1 \dots \omega_n H) - S_{n+1}(\omega_1 \dots \omega_n T)}. \quad (1.2.17)$$

The numbers $\tilde{p}$ and $\tilde{q}$ appearing in (1.2.16) are the *risk-neutral probabilities* given by

$$\tilde{p} = \frac{1+r-d}{u-d}, \quad \tilde{q} = \frac{u-1-r}{u-d}. \quad (1.2.15)$$

These risk-neutral probabilities are positive because of (1.1.2) and sum to 1. They have the property that, at any time, the price of the stock is the discounted risk-neutral average of its two possible prices at the next time:

$$S_n(\omega_1 \dots \omega_n) = \frac{1}{1+r}\big[\tilde{p}S_{n+1}(\omega_1 \dots \omega_n H) + \tilde{q}S_{n+1}(\omega_1 \dots \omega_n T)\big].$$

In other words, under the risk-neutral probabilities, the mean rate of return for the stock is r, the same as the rate of return for the money market. Therefore, if these probabilities actually governed the coin tossing (in fact, they do not), then an agent trading in the money market account and stock would have before him two opportunities, both of which provide the same mean rate of return. Consequently, no matter how he invests, the mean rate of return for his portfolio would also be r. In particular, if it is time $N-1$ and he wants his portfolio value to be $V_N(\omega_1 \dots \omega_N)$ at time N, then at time $N-1$ his portfolio value must be

$$\frac{1}{1+r}\big[\tilde{p}V_N(\omega_1 \dots \omega_{N-1}H) + \tilde{q}V_N(\omega_1 \dots \omega_{N-1}T)\big].$$

This is the right-hand side of (1.2.16) with $n = N-1$, and repeated application of this argument yields (1.2.16) for all values of n.

The explanation of (1.2.16) above was given under a condition contrary to fact, namely that $\tilde{p}$ and $\tilde{q}$ govern the coin tossing. One can ask whether such an argument can result in a valid conclusion. It does result in a valid conclusion for the following reason. When hedging a short position in a derivative security, we want the hedge to give us a portfolio that agrees with the payoff of the derivative security *regardless of the coin tossing.* In other words, the hedge must work *on all stock price paths.* If a path is possible (i.e., has positive probability), we want the hedge to work along that path. The actual value of the probability is irrelevant. We find these hedges by solving a system of equations along the paths, a system of the form (1.2.2)–(1.2.3), (1.2.5)–(1.2.8). There are no probabilities in this system. Introducing the risk-neutral probabilities allows us to argue as above and find a solution to the system. Introducing any other probabilities would not allow such an argument because only the risk-neutral probabilities allow us to state that no matter how the agent invests, the mean rate of return for his portfolio is r. The risk-neutral probabilities provide a shortcut to solving the system of equations. The actual probabilities are no help in solving this system. Under the actual probabilities, the mean rate of return for a portfolio depends on the portfolio, and when we are trying to solve the system of equations, we do not know which portfolio we should use.

Alternatively, one can explain (1.2.16) without recourse to any discussion of probability. This was the approach taken in the proof of Theorem 1.2.2. The numbers $\tilde{p}$ and $\tilde{q}$ were used in that proof, but they were not regarded as probabilities, just numbers defined by the formula (1.2.15).

1.5 Notes

No-arbitrage pricing is implicit in the work of Black and Scholes [5], but its first explicit development is provided by Merton [34], who began with the axiom of no-arbitrage and obtained a surprising number of conclusions. No arbitrage pricing was fully developed in continuous-time models by Harrison and Kreps [17] and Harrison and Pliska [18]. These authors introduced martingales (Sections 2.4 in this text and Section 2.3 in Volume II) and risk-neutral pricing. The binomial model is due to Cox, Ross, Rubinstein [11]; a good reference is [12]. The binomial model is useful in its own right, and as Cox et al. showed, one can rederive the Black-Scholes-Merton formula as a limit of the binomial model (see Theorem 3.2.2 in Chapter 3 of Volume II for the log-normality of the stock price obtained in the limit of the binomial model.)

1.6 Exercises

Exercise 1.1. Assume in the one-period binomial market of Section 1.1 that both H and T have positive probability of occurring. Show that condition (1.1.2) precludes arbitrage. In other words, show that if $X_0 = 0$ and

$$X_1 = \Delta_0 S_1 + (1+r)(X_0 - \Delta_0 S_0),$$

then we cannot have X_1 strictly positive with positive probability unless X_1 is strictly negative with positive probability as well, and this is the case regardless of the choice of the number Δ_0.

Exercise 1.2. Suppose in the situation of Example 1.1.1 that the option sells for 1.20 at time zero. Consider an agent who begins with wealth $X_0 = 0$ and at time zero buys Δ_0 shares of stock and Γ_0 options. The numbers Δ_0 and Γ_0 can be either positive or negative or zero. This leaves the agent with a cash position of $-4\Delta_0 - 1.20\Gamma_0$. If this is positive, it is invested in the money market; if it is negative, it represents money borrowed from the money market. At time one, the value of the agent's portfolio of stock, option, and money market assets is

$$X_1 = \Delta_0 S_1 + \Gamma_0 (S_1 - 5)^+ - \frac{5}{4}(4\Delta_0 + 1.20\Gamma_0).$$

Assume that both H and T have positive probability of occurring. Show that if there is a positive probability that X_1 is positive, then there is a positive probability that X_1 is negative. In other words, one cannot find an arbitrage when the time-zero price of the option is 1.20.

Exercise 1.3. In the one-period binomial model of Section 1.1, suppose we want to determine the price at time zero of the derivative security $V_1 = S_1$ (i.e., the derivative security pays off the stock price.) (This can be regarded as a European call with strike price $K = 0$). What is the time-zero price V_0 given by the risk-neutral pricing formula (1.1.10)?

Exercise 1.4. In the proof of Theorem 1.2.2, show under the induction hypothesis that

$$X_{n+1}(\omega_1\omega_2\ldots\omega_n T) = V_{n+1}(\omega_1\omega_2\ldots\omega_n T).$$

Exercise 1.5. In Example 1.2.4, we considered an agent who sold the lookback option for $V_0 = 1.376$ and bought $\Delta_0 = 0.1733$ shares of stock at time zero. At time one, if the stock goes up, she has a portfolio valued at $V_1(H) = 2.24$. Assume that she now takes a position of $\Delta_1(H) = \frac{V_2(HH)-V_2(HT)}{S_2(HH)-S_2(HT)}$ in the stock. Show that, at time two, if the stock goes up again, she will have a portfolio valued at $V_2(HH) = 3.20$, whereas if the stock goes down, her portfolio will be worth $V_2(HT) = 2.40$. Finally, under the assumption that the stock goes up in the first period and down in the second period, assume the agent takes a position of $\Delta_2(HT) = \frac{V_3(HTH)-V_3(HTT)}{S_3(HTH)-S_3(HTT)}$ in the stock. Show that, at time three, if the stock goes up in the third period, she will have a portfolio valued at $V_3(HTH) = 0$, whereas if the stock goes down, her portfolio will be worth $V_3(HTT) = 6$. In other words, she has hedged her short position in the option.

Exercise 1.6 (Hedging a long position-one period). Consider a bank that has a long position in the European call written on the stock price in Figure 1.1.2. The call expires at time one and has strike price $K = 5$. In Section 1.1, we determined the time-zero price of this call to be $V_0 = 1.20$. At time zero, the bank owns this option, which ties up capital $V_0 = 1.20$. The bank wants to earn the interest rate 25% on this capital until time one (i.e., without investing any more money, and regardless of how the coin tossing turns out, the bank wants to have

$$\frac{5}{4} \cdot 1.20 = 1.50$$

at time one, after collecting the payoff from the option (if any) at time one). Specify how the bank's trader should invest in the stock and money markets to accomplish this.

Exercise 1.7 (Hedging a long position-multiple periods). Consider a bank that has a long position in the lookback option of Example 1.2.4. The bank intends to hold this option until expiration and receive the payoff V_3. At time zero, the bank has capital $V_0 = 1.376$ tied up in the option and wants to earn the interest rate of 25% on this capital until time three (i.e., without investing any more money, and regardless of how the coin tossing turns out, the bank wants to have

$$\left(\frac{5}{4}\right)^3 \cdot 1.376 = 2.6875$$

at time three, after collecting the payoff from the lookback option at time three). Specify how the bank's trader should invest in the stock and the money market account to accomplish this.

Exercise 1.8 (Asian option). Consider the three-period model of Example 1.2.1, with $S_0 = 4$, $u = 2$, $d = \frac{1}{2}$, and take the interest rate $r = \frac{1}{4}$, so that $\tilde{p} = \tilde{q} = \frac{1}{2}$. For $n = 0, 1, 2, 3$, define $Y_n = \sum_{k=0}^{n} S_k$ to be the sum of the stock prices between times zero and n. Consider an *Asian call option* that expires at time three and has strike $K = 4$ (i.e., whose payoff at time three is $\left(\frac{1}{4}Y_3 - 4\right)^+$). This is like a European call, except the payoff of the option is based on the average stock price rather than the final stock price. Let $v_n(s, y)$ denote the price of this option at time n if $S_n = s$ and $Y_n = y$. In particular, $v_3(s, y) = \left(\frac{1}{4}y - 4\right)^+$.

(i) Develop an algorithm for computing v_n recursively. In particular, write a formula for v_n in terms of v_{n+1}.
(ii) Apply the algorithm developed in (i) to compute $v_0(4, 4)$, the price of the Asian option at time zero.
(iii) Provide a formula for $\delta_n(s, y)$, the number of shares of stock that should be held by the replicating portfolio at time n if $S_n = s$ and $Y_n = y$.

Exercise 1.9 (Stochastic volatility, random interest rate). Consider a binomial pricing model, but at each time $n \geq 1$, the "up factor" $u_n(\omega_1\omega_2 \ldots \omega_n)$, the "down factor" $d_n(\omega_1\omega_2 \ldots \omega_n)$, and the interest rate $r_n(\omega_1\omega_2 \ldots \omega_n)$ are allowed to depend on n and on the first n coin tosses $\omega_1\omega_2 \ldots \omega_n$. The initial up factor u_0, the initial down factor d_0, and the initial interest rate r_0 are not random. More specifically, the stock price at time one is given by

$$S_1(\omega_1) = \begin{cases} u_0 S_0 \text{ if } \omega_1 = H, \\ d_0 S_0 \text{ if } \omega_1 = T, \end{cases}$$

and, for $n \geq 1$, the stock price at time $n + 1$ is given by

$$S_{n+1}(\omega_1\omega_2 \ldots \omega_n\omega_{n+1}) = \begin{cases} u_n(\omega_1\omega_2 \ldots \omega_n) S_n(\omega_1\omega_2 \ldots \omega_n) \text{ if } \omega_{n+1} = H, \\ d_n(\omega_1\omega_2 \ldots \omega_n) S_n(\omega_1\omega_2 \ldots \omega_n) \text{ if } \omega_{n+1} = T. \end{cases}$$

One dollar invested in or borrowed from the money market at time zero grows to an investment or debt of $1+r_0$ at time one, and, for $n \geq 1$, one dollar invested in or borrowed from the money market at time n grows to an investment or debt of $1+r_n(\omega_1\omega_2\ldots\omega_n)$ at time $n+1$. We assume that for each n and for all $\omega_1\omega_2\ldots\omega_n$, the no-arbitrage condition

$$0 < d_n(\omega_1\omega_2\ldots\omega_n) < 1+r_n(\omega_1\omega_2\ldots\omega_n) < u_n(\omega_1\omega_2\ldots\omega_n)$$

holds. We also assume that $0 < d_0 < 1+r_0 < u_0$.

(i) Let N be a positive integer. In the model just described, provide an algorithm for determining the price at time zero for a derivative security that at time N pays off a random amount V_N depending on the result of the first N coin tosses.

(ii) Provide a formula for the number of shares of stock that should be held at each time n $(0 \leq n \leq N-1)$ by a portfolio that replicates the derivative security V_N.

(iii) Suppose the initial stock price is $S_0 = 80$, with each head the stock price increases by 10, and with each tail the stock price decreases by 10. In other words, $S_1(H) = 90$, $S_1(T) = 70$, $S_2(HH) = 100$, etc. Assume the interest rate is always zero. Consider a European call with strike price 80, expiring at time five. What is the price of this call at time zero?

2

Probability Theory on Coin Toss Space

2.1 Finite Probability Spaces

A finite probability space is used to model a situation in which a random experiment with finitely many possible outcomes is conducted. In the context of the binomial model of the previous chapter, we tossed a coin a finite number of times. If, for example, we toss the coin three times, the set of all possible outcomes is

$$\Omega = \{HHH, HHT, HTH, HTT, THH, THT, TTH, TTT\}. \tag{2.1.1}$$

Suppose that on each toss the probability of a head (either actual or risk-neutral) is p and the probability of a tail is $q = 1-p$. We assume the tosses are independent, and so the probabilities of the individual elements ω (sequences of three tosses $\omega = \omega_1\omega_2\omega_3$) in Ω are

$$\begin{aligned} &\mathbb{P}(HHH) = p^3,\ \mathbb{P}(HHT) = p^2q,\ \mathbb{P}(HTH) = p^2q,\ \mathbb{P}(HTT) = pq^2,\\ &\mathbb{P}(THH) = p^2q,\ \mathbb{P}(THT) = pq^2,\ \mathbb{P}(TTH) = pq^2,\ \ \mathbb{P}(TTT) = q^3. \end{aligned} \tag{2.1.2}$$

The subsets of Ω are called *events*, and these can often be described in words as well as in symbols. For example, the event

$$\begin{aligned} \text{“The first toss is a head”} &= \{\omega \in \Omega; \omega_1 = H\}\\ &= \{HHH, HHT, HTH, HTT\} \end{aligned}$$

has, as indicated, descriptions in both words and symbols. We determine the probability of an event by summing the probabilities of the elements in the event, i.e.,

$$\begin{aligned} \mathbb{P}(\text{First toss is a head}) &= \mathbb{P}(HHH) + \mathbb{P}(HHT) + \mathbb{P}(HTH) + \mathbb{P}(HTT)\\ &= (p^3 + p^2q) + (p^2q + pq^2)\\ &= p^2(p+q) + pq(p+q) \end{aligned}$$

$$
\begin{aligned}
&= p^2 + pq \\
&= p(p+q) \\
&= p. \qquad (2.1.3)
\end{aligned}
$$

Thus, the mathematics agrees with our intuition.

With mathematical models, it is easy to substitute our intuition for the mathematics, but this can lead to trouble. We should instead verify that the mathematics and our intuition agree; otherwise, either our intuition is wrong or our model is inadequate. If our intuition and the mathematics of a model do not agree, we should seek a reconciliation before proceeding. In the case of (2.1.3), we set out to build a model in which the probability of a head on each toss is p, we proposed doing this by defining the probabilities of the elements of Ω by (2.1.2), and we further defined the probability of an event (subset of Ω) to be the sum of the probabilities of the elements in the event. These definitions force us to carry out the computation (2.1.3) as we have done, and we need to do this computation in order to check that it gets the expected answer. Otherwise, we would have to rethink our mathematical model for the coin tossing.

We generalize slightly the situation just described, first by allowing Ω to be any finite set, and second by allowing some elements in Ω to have probability zero. These generalizations lead to the following definition.

Definition 2.1.1. *A* finite probability space *consists of a* sample space Ω *and a* probability measure $\mathbb{P}$. *The sample space Ω is a nonempty finite set and the probability measure $\mathbb{P}$ is a function that assigns to each element ω of Ω a number in* $[0,1]$ *so that*

$$\sum_{\omega \in \Omega} \mathbb{P}(\omega) = 1. \qquad (2.1.4)$$

An event *is a subset of Ω, and we define the probability of an event A to be*

$$\mathbb{P}(A) = \sum_{\omega \in A} \mathbb{P}(\omega). \qquad (2.1.5)$$

As mentioned before, this is a model for some random experiment. The set Ω is the set of all possible outcomes of the experiment, $\mathbb{P}(\omega)$ is the probability that the particular outcome ω occurs, and $\mathbb{P}(A)$ is the probability that the outcome that occurs is in the set A. If $\mathbb{P}(A) = 0$, then the outcome of the experiment is sure not to be in A; if $\mathbb{P}(A) = 1$, then the outcome is sure to be in A. Because of (2.1.4), we have the equation

$$\mathbb{P}(\Omega) = 1, \qquad (2.1.6)$$

i.e., the outcome that occurs is sure to be in the set Ω. Because $\mathbb{P}(\omega)$ can be zero for some values of ω, we are permitted to put in Ω even some outcomes of the experiment that are sure not to occur. It is clear from (2.1.5) that if A and B are disjoint subsets of Ω, then

$$\mathbb{P}(A \cup B) = \mathbb{P}(A) + \mathbb{P}(B). \qquad (2.1.7)$$

2.2 Random Variables, Distributions, and Expectations

A random experiment generally generates numerical data. This gives rise to the concept of a random variable.

Definition 2.2.1. *Let* $(\Omega, \mathbb{P})$ *be a finite probability space. A* random variable *is a real-valued function defined on* Ω. *(We sometimes also permit a random variable to take the values* $+\infty$ *and* $-\infty$*.)*

Example 2.2.2 (Stock prices). Recall the space Ω of three independent coin-tosses (2.1.1). As in Figure 1.2.2 of Chapter 1, let us define stock prices by the formulas

$$S_0(\omega_1\omega_2\omega_3) = 4 \text{ for all } \omega_1\omega_2\omega_3 \in \Omega_3,$$

$$S_1(\omega_1\omega_2\omega_3) = \begin{cases} 8 & \text{if } \omega_1 = H, \\ 2 & \text{if } \omega_1 = T, \end{cases}$$

$$S_2(\omega_1\omega_2\omega_3) = \begin{cases} 16 & \text{if } \omega_1 = \omega_2 = H, \\ 4 & \text{if } \omega_1 \neq \omega_2, \\ 1 & \text{if } \omega_1 = \omega_2 = T, \end{cases}$$

$$S_3(\omega_1\omega_2\omega_3) = \begin{cases} 32 & \text{if } \omega_1 = \omega_2 = \omega_3 = H, \\ 8 & \text{if there are two heads and one tail,} \\ 2 & \text{if there is one head and two tails,} \\ .50 & \text{if } \omega_1 = \omega_2 = \omega_3 = T. \end{cases}$$

Here we have written the arguments of S_0, S_1, S_2, and S_3 as $\omega_1\omega_2\omega_3$, even though some of these random variables do not depend on all the coin tosses. In particular, S_0 is actually not random because it takes the value 4, regardless of how the coin tosses turn out; such a random variable is sometimes called a *degenerate random variable.* □

It is customary to write the argument of random variables as ω, even when ω is a sequence such as $\omega = \omega_1\omega_2\omega_3$. We shall use these two notations interchangeably. It is even more common to write random variables without any arguments; we shall switch to that practice presently, writing S_3, for example, rather than $S_3(\omega_1\omega_2\omega_3)$ or $S_3(\omega)$.

According to Definition 2.2.1, a random variable is a function that maps a sample space Ω to the real numbers. The *distribution* of a random variable is a specification of the probabilities that the random variable takes various values. *A random variable is not a distribution*, and *a distribution is not a random variable.* This is an important point when we later switch between the actual probability measure, which one would estimate from historical data, and the risk-neutral probability measure. The change of measure will change distributions of random variables but not the random variables themselves. We make this distinction clear with the following example.

Example 2.2.3. Toss a coin three times, so the set of possible outcomes is

$$\Omega = \{HHH, HHT, HTH, HTT, THH, THT, TTH, TTT\}.$$

Define the random variables

$$X = \text{Total number of heads}, \qquad Y = \text{Total number of tails}.$$

In symbols,

$$\begin{aligned} X(HHH) &= 3, \\ X(HHT) = X(HTH) = X(THH) &= 2, \\ X(HTT) = X(THT) = X(TTH) &= 1, \\ X(TTT) &= 0, \end{aligned}$$

$$\begin{aligned} Y(TTT) &= 3, \\ Y(TTH) = Y(THT) = Y(HTT) &= 2, \\ Y(THH) = Y(HTH) = Y(HHT) &= 1, \\ Y(HHH) &= 0. \end{aligned}$$

We do not need to know probabilities of various outcomes in order to specify these random variables. However, once we specify a probability measure on Ω, we can determine the distributions of X and Y. For example, if we specify the probability measure $\widetilde{\mathbb{P}}$ under which the probability of head on each toss is $\frac{1}{2}$ and the probability of each element in Ω is $\frac{1}{8}$, then

$$\widetilde{\mathbb{P}}\{\omega \in \Omega; X(\omega) = 0\} = \widetilde{\mathbb{P}}\{TTT\} = \frac{1}{8},$$

$$\widetilde{\mathbb{P}}\{\omega \in \Omega; X(\omega) = 1\} = \widetilde{\mathbb{P}}\{HTT, THT, TTH\} = \frac{3}{8},$$

$$\widetilde{\mathbb{P}}\{\omega \in \Omega; X(\omega) = 2\} = \widetilde{\mathbb{P}}\{HHT, HTH, THH\} = \frac{3}{8},$$

$$\widetilde{\mathbb{P}}\{\omega \in \Omega; X(\omega) = 3\} = \widetilde{\mathbb{P}}\{HHH\} = \frac{1}{8}.$$

We shorten the cumbersome notation $\widetilde{\mathbb{P}}\{\omega \in \Omega; X(\omega) = j\}$ to simply $\widetilde{\mathbb{P}}\{X = j\}$. It is helpful to remember, however, that the notation $\widetilde{\mathbb{P}}\{X = j\}$ refers to the probability of a subset of Ω, the set of elements ω for which $X(\omega) = j$. Under $\widetilde{\mathbb{P}}$, the probability that X takes the four values 0, 1, 2, and 3 are

$$\widetilde{\mathbb{P}}\{X = 0\} = \frac{1}{8}, \ \widetilde{\mathbb{P}}\{X = 1\} = \frac{3}{8},$$
$$\widetilde{\mathbb{P}}\{X = 2\} = \frac{3}{8}, \ \widetilde{\mathbb{P}}\{X = 3\} = \frac{1}{8}.$$

This table of probabilities where X takes its various values records the *distribution* of X under $\widetilde{\mathbb{P}}$.

The random variable Y is different from X because it counts tails rather than heads. However, under $\widetilde{\mathbb{P}}$, the distribution of Y is the same as the distribution of X:

$$\widetilde{\mathbb{P}}\{Y=0\}=\frac{1}{8},\ \widetilde{\mathbb{P}}\{Y=1\}=\frac{3}{8},$$
$$\widetilde{\mathbb{P}}\{Y=2\}=\frac{3}{8},\ \widetilde{\mathbb{P}}\{Y=3\}=\frac{1}{8}.$$

The point here is that the random variable is a function defined on Ω, whereas its distribution is a tabulation of probabilities that the random variable takes various values. A random variable is not a distribution.

Suppose, moreover, that we choose a probability measure $\mathbb{P}$ for Ω that corresponds to a $\frac{2}{3}$ probability of head on each toss and a $\frac{1}{3}$ probability of tail. Then

$$\mathbb{P}\{X=0\}=\frac{1}{27},\ \mathbb{P}\{X=1\}=\frac{6}{27},$$
$$\mathbb{P}\{X=2\}=\frac{12}{27},\ \mathbb{P}\{X=3\}=\frac{8}{27}.$$

The random variable X has a different distribution under $\mathbb{P}$ than under $\widetilde{\mathbb{P}}$. It is the same random variable, counting the total number of heads, regardless of the probability measure used to determine its distribution. This is the situation we encounter later when we consider an asset price under both the actual and the risk-neutral probability measures.

Incidentally, although they have the same distribution under $\widetilde{\mathbb{P}}$, the random variables X and Y have different distributions under $\mathbb{P}$. Indeed,

$$\mathbb{P}\{Y=0\}=\frac{8}{27},\quad \mathbb{P}\{Y=1\}=\frac{12}{27},$$
$$\mathbb{P}\{Y=2\}=\frac{6}{27},\quad \mathbb{P}\{Y=3\}=\frac{1}{27}. \qquad \square$$

Definition 2.2.4. *Let X be a random variable defined on a finite probability space $(\Omega,\mathbb{P})$. The* expectation *(or* expected value*) of X is defined to be*

$$\mathbb{E}X=\sum_{\omega\in\Omega}X(\omega)\mathbb{P}(\omega).$$

When we compute the expectation using the risk-neutral probability measure $\widetilde{\mathbb{P}}$, we use the notation

$$\widetilde{\mathbb{E}}X=\sum_{\omega\in\Omega}X(\omega)\widetilde{\mathbb{P}}(\omega).$$

The variance *of X is*

$$\mathrm{Var}(X)=\mathbb{E}\Big[(X-\mathbb{E}X)^2\Big].$$

It is clear from its definition that expectation is linear: if X and Y are random variables and c_1 and c_2 are constants, then

$$\mathbb{E}(c_1X + c_2Y) = c_1\mathbb{E}X + c_2\mathbb{E}Y.$$

In particular, if $\ell(x) = ax+b$ is a linear function of a dummy variable x (a and b are constants), then $\mathbb{E}[\ell(X)] = \ell(\mathbb{E}X)$. When dealing with convex functions, we have the following inequality.

Theorem 2.2.5 (Jensen's inequality). *Let X be a random variable on a finite probability space, and let $\varphi(x)$ be a convex function of a dummy variable x. Then*

$$\mathbb{E}[\varphi(X)] \geq \varphi(\mathbb{E}X).$$

PROOF: We first argue that a convex function is the maximum of all linear functions that lie below it; i.e., for every $x \in \mathbb{R}$,

$$\varphi(x) = \max\{\ell(x); \ell \text{ is linear and } \ell(y) \leq \varphi(y) \text{ for all } y \in \mathbb{R}\}. \tag{2.2.1}$$

Since we are only considering linear functions that lie below φ, it is clear that

$$\varphi(x) \geq \max\{\ell(x); \ell \text{ is linear and } \ell(y) \leq \varphi(y) \text{ for all } y \in \mathbb{R}\}.$$

On the other hand, let x be an arbitrary point in $\mathbb{R}$. Because φ is convex, there is always a linear function ℓ that lies below φ and for which $\varphi(x) = \ell(x)$ for this particular x. This is called a *support line of φ at x* (see Figure 2.2.1). Therefore,

$$\varphi(x) \leq \max\{\ell(x); \ell \text{ is linear and } \ell(y) \leq \varphi(y) \text{ for all } y \in \mathbb{R}\}.$$

This establishes (2.2.1). Now let ℓ be a linear function lying below φ. We have

$$\mathbb{E}[\varphi(X)] \geq \mathbb{E}[\ell(X)] = \ell(\mathbb{E}X).$$

Since this inequality holds for every linear function ℓ lying below φ, we may take the maximum on the right-hand side over all such ℓ and obtain

$$\begin{aligned}\mathbb{E}[\varphi(X)] &\geq \max\{\ell(\mathbb{E}X); \ell \text{ is linear and } \ell(y) \leq \varphi(y) \text{ for all } y \in \mathbb{R}\}\\ &= \varphi(\mathbb{E}X).\end{aligned}$$

□

One consequence of Jensen's inequality is that

$$\mathbb{E}\left[X^2\right] \geq (\mathbb{E}X)^2.$$

We can also obtain this particular consequence of Jensen's inequality from the formula

$$0 \leq \mathbb{E}\Big[(X - \mathbb{E}X)^2\Big] = \mathbb{E}\Big[X^2 - 2X\mathbb{E}X + (\mathbb{E}X)^2\Big] = \mathbb{E}[X^2] - (\mathbb{E}X)^2.$$

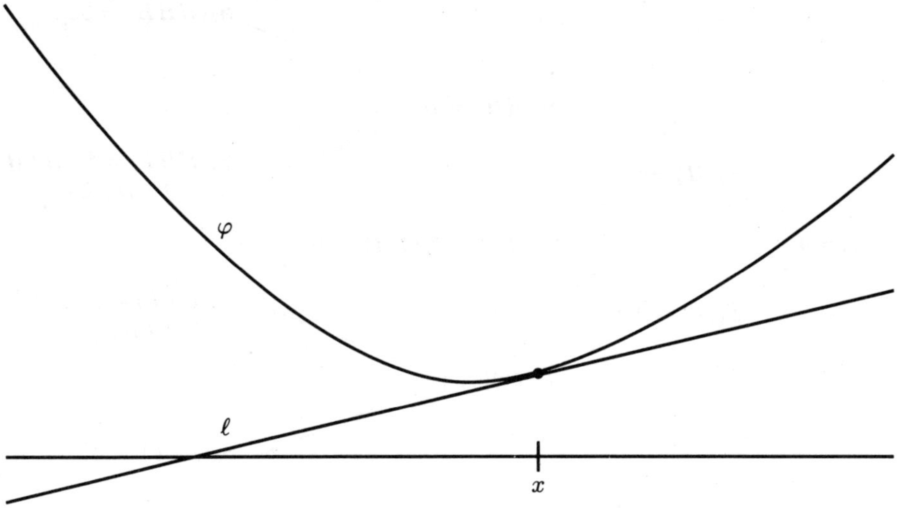

Fig. 2.2.1. Support line of φ at x.

2.3 Conditional Expectations

In the binomial pricing model of Chapter 1, we chose risk-neutral probabilities $\tilde{p}$ and $\tilde{q}$ by the formula (1.1.8), which we repeat here:

$$\tilde{p} = \frac{1+r-d}{u-d}, \quad \tilde{q} = \frac{u-1-r}{u-d}. \tag{2.3.1}$$

It is easily checked that these probabilities satisfy the equation

$$\frac{\tilde{p}u + \tilde{q}d}{1+r} = 1. \tag{2.3.2}$$

Consequently, at every time n and for every sequence of coin tosses $\omega_1 \dots \omega_n$, we have

$$S_n(\omega_1 \dots \omega_n) = \frac{1}{1+r}\Big[\tilde{p}S_{n+1}(\omega_1 \dots \omega_n H) + \tilde{q}S_{n+1}(\omega_1 \dots \omega_n T)\Big] \tag{2.3.3}$$

(i.e., the stock price at time n is the discounted weighted average of the two possible stock prices at time $n+1$, where $\tilde{p}$ and $\tilde{q}$ are the weights used in the averaging). To simplify notation, we define

$$\widetilde{\mathbb{E}}_n[S_{n+1}](\omega_1 \dots \omega_n) = \tilde{p}S_{n+1}(\omega_1 \dots \omega_n H) + \tilde{q}S_{n+1}(\omega_1 \dots \omega_n T) \tag{2.3.4}$$

so that we may rewrite (2.3.3) as

$$S_n = \frac{1}{1+r}\widetilde{\mathbb{E}}_n[S_{n+1}], \tag{2.3.5}$$

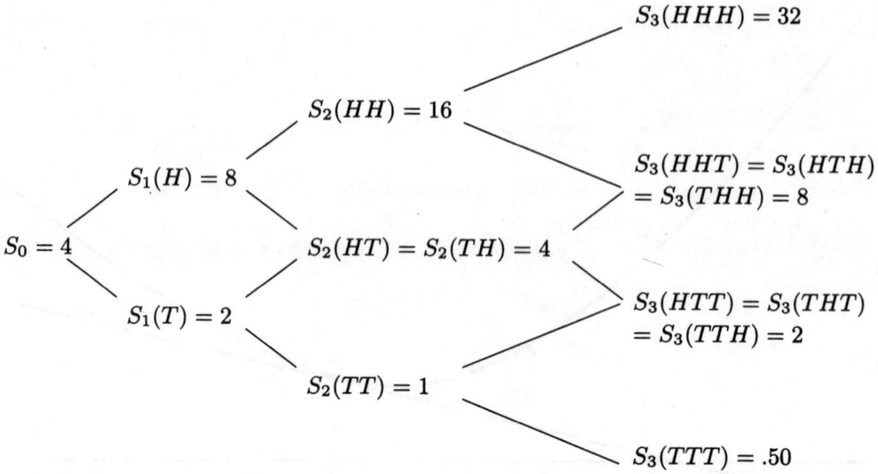

Fig. 2.3.1. A three-period model.

and we call $\widetilde{\mathbb{E}}_n[S_{n+1}]$ the *conditional expectation of* S_{n+1} *based on the information at time* n. The conditional expectation can be regarded as an estimate of the value of S_{n+1} based on knowledge of the first n coin tosses.

For example, in Figure 2.3.1 and using the risk-neutral probabilities $\tilde{p} = \tilde{q} = \frac{1}{2}$, we have $\widetilde{\mathbb{E}}_1[S_2](H) = 10$ and $\widetilde{\mathbb{E}}_1[S_2](T) = 2.50$. When we write simply $\widetilde{\mathbb{E}}_1[S_2]$ without specifying whether the first coin toss results in head or tail, we have a quantity whose value, not known at time zero, will be determined by the random experiment of coin tossing. According to Definition 2.2.1, such a quantity is a random variable.

More generally, whenever X is a random variable depending on the first N coin tosses, we can estimate X based on information available at an earlier time $n \leq N$. The following definition generalizes (2.3.4).

Definition 2.3.1. *Let n satisfy $1 \leq n \leq N$, and let $\omega_1 \dots \omega_n$ be given and, for the moment, fixed. There are 2^{N-n} possible continuations $\omega_{n+1} \dots \omega_N$ of the sequence fixed $\omega_1 \dots \omega_n$. Denote by $\#H(\omega_{n+1} \dots \omega_N)$ the number of heads in the continuation $\omega_{n+1} \dots \omega_N$ and by $\#T(\omega_{n+1} \dots \omega_N)$ the number of tails. We define*

$$\widetilde{\mathbb{E}}_n[X](\omega_1 \dots \omega_n) = \sum_{\omega_{n+1} \dots \omega_N} \tilde{p}^{\#H(\omega_{n+1} \dots \omega_N)} \tilde{q}^{\#T(\omega_{n+1} \dots \omega_N)} X(\omega_1 \dots \omega_n \omega_{n+1} \dots \omega_N) \quad (2.3.6)$$

and call $\widetilde{\mathbb{E}}_n[X]$ the conditional expectation of X based on the information at time n.

Based on what we know at time zero, the conditional expectation $\widetilde{\mathbb{E}}_n[X]$ is random in the sense that its value depends on the first n coin tosses, which we do not know until time n. For example, in Figure 2.3.1 and using $\tilde{p} = \tilde{q} = \frac{1}{2}$, we obtain

$$\widetilde{\mathbb{E}}_1[S_3](H) = 12.50, \quad \widetilde{\mathbb{E}}_1[S_3](T) = 3.125,$$

so $\widetilde{\mathbb{E}}_1[S_3]$ is a random variable.

Definition 2.3.1 continued *The two extreme cases of conditioning are* $\widetilde{\mathbb{E}}_0[X]$, *the conditional expectation of* X *based on no information, which we define by*

$$\widetilde{\mathbb{E}}_0[X] = \widetilde{\mathbb{E}}X, \tag{2.3.7}$$

and $\widetilde{\mathbb{E}}_N[X]$, *the conditional expectation of* X *based on knowledge of all* N *coin tosses, which we define by*

$$\widetilde{\mathbb{E}}_N[X] = X. \tag{2.3.8}$$

The conditional expectations above have been computed using the risk-neutral probabilities $\tilde{p}$ and $\tilde{q}$. This is indicated by the $\sim$ appearing in the notation $\widetilde{\mathbb{E}}_n$. Of course, conditional expectations can also be computed using the actual probabilities p and q, and these will be denoted by $\mathbb{E}_n$.

Regarded as random variables, conditional expectations have five fundamental properties, which we will use extensively. These are listed in the following theorem. We state them for conditional expectations computed under the actual probabilities, and the analogous results hold for conditional expectations computed under the risk-neutral probabilities.

Theorem 2.3.2 (Fundamental properties of conditional expectations). *Let* N *be a positive integer, and let* X *and* Y *be random variables depending on the first* N *coin tosses. Let* $0 \leq n \leq N$ *be given. The following properties hold.*

(i) ***Linearity of conditional expectations****. For all constants* c_1 *and* c_2, *we have*

$$\mathbb{E}_n[c_1X + c_2Y] = c_1\mathbb{E}_n[X] + c_2\mathbb{E}_n[Y].$$

(ii) ***Taking out what is known****. If* X *actually depends only on the first* n *coin tosses, then*

$$\mathbb{E}_n[XY] = X \cdot \mathbb{E}_n[Y].$$

(iii) ***Iterated conditioning****. If* $0 \leq n \leq m \leq N$, *then*

$$\mathbb{E}_n\Big[\mathbb{E}_m[X]\Big] = \mathbb{E}_n[X].$$

In particular, $\mathbb{E}\Big[\mathbb{E}_m[X]\Big] = \mathbb{E}X$.

(iv) ***Independence****. If* X *depends only on tosses* $n+1$ *through* N, *then*

$$\mathbb{E}_n[X] = \mathbb{E}X.$$

(v) ***Conditional Jensen's inequality.*** *If $\varphi(x)$ is a convex function of the dummy variable x, then*

$$\mathbb{E}_n[\varphi(X)] \geq \varphi(\mathbb{E}_n[X]).$$

The proof of Theorem 2.3.2 is provided in the appendix. We illustrate the first four properties of the theorem with examples based on Figure 2.3.1 using the probabilities $p = \frac{2}{3}$, $q = \frac{1}{3}$. The fifth property, the conditional Jensen's inequality, follows from linearity of conditional expectations in the same way that Jensen's inequality for expectations follows from linearity of expectations (see the proof of Theorem 2.2.5).

Example 2.3.3 (Linearity of conditional expectations). With $p = \frac{2}{3}$ and $q = \frac{1}{3}$ in Figure 2.3.1, we compute

$$\mathbb{E}_1[S_2](H) = \frac{2}{3} \cdot 16 + \frac{1}{3} \cdot 4 \;=\; 12,$$
$$\mathbb{E}_1[S_3](H) = \frac{4}{9} \cdot 32 + \frac{2}{9} \cdot 8 + \frac{2}{9} \cdot 8 + \frac{1}{9} \cdot 2 \;=\; 18,$$

and consequently $\mathbb{E}_1[S_2](H) + \mathbb{E}_1[S_3](H) = 12 + 18 = 30$. But also

$$\mathbb{E}_1[S_2 + S_3](H) = \frac{4}{9}(16+32) + \frac{2}{9}(16+8) + \frac{2}{9}(4+8) + \frac{1}{9}(4+2) = 30.$$

A similar computation shows that

$$\mathbb{E}_1[S_2 + S_3](T) = 7.50 = \mathbb{E}_1[S_2](T) + \mathbb{E}_1[S_3](T).$$

In conclusion, regardless of the outcome of the first coin toss,

$$\mathbb{E}_1[S_2 + S_3] = \mathbb{E}_1[S_2] + E_1[S_3].$$

Example 2.3.4 (Taking out what is known). We first recall from Example 2.3.3 that

$$\mathbb{E}_1[S_2](H) = \frac{2}{3} \cdot 16 + \frac{1}{3} \cdot 4 = 12.$$

If we now want to estimate the product $S_1 S_2$ based on the information at time one, we can factor out the S_1, as seen by the following computation:

$$\mathbb{E}_1[S_1 S_2](H) = \frac{2}{3} \cdot 128 + \frac{1}{3} \cdot 32 = 96 = 8 \cdot 12 = S_1(H)\mathbb{E}_1[S_2](H).$$

A similar computation shows that

$$E_1[S_1 S_2](T) = 6 = S_1(T)\mathbb{E}_1[S_2](T).$$

In conclusion, regardless of the outcome of the first toss,

$$\mathbb{E}_1[S_1 S_2] = S_1 \mathbb{E}_1[S_2].$$

Example 2.3.5 (Iterated conditioning). We first estimate S_3 based on the information at time two:

$$\begin{aligned}
\mathbb{E}_2[S_3](HH) &= \frac{2}{3}\cdot 32 + \frac{1}{3}\cdot 8 = 24,\\
\mathbb{E}_2[S_3](HT) &= \frac{2}{3}\cdot 8 + \frac{1}{3}\cdot 2 = 6,\\
\mathbb{E}_2[S_3](TH) &= \frac{2}{3}\cdot 8 + \frac{1}{3}\cdot 2 = 6,\\
\mathbb{E}_2[S_3](TT) &= \frac{2}{3}\cdot 2 + \frac{1}{3}\cdot \frac{1}{2} = 1.50.
\end{aligned}$$

We now estimate the estimate, based on the information at time one:

$$\begin{aligned}
\mathbb{E}_1\Big[\mathbb{E}_2[S_3]\Big](H) &= \frac{2}{3}\cdot \mathbb{E}_2[S_3](HH) + \frac{1}{3}\cdot \mathbb{E}_2[S_3](HT)\\
&= \frac{2}{3}\cdot 24 + \frac{1}{3}\cdot 6 \;=\; 18,\\
\mathbb{E}_1\Big[\mathbb{E}_2[S_3]\Big](T) &= \frac{2}{3}\cdot \mathbb{E}_2[S_3](TH) + \frac{1}{3}\mathbb{E}_2[S_3](TT)\\
&= \frac{2}{3}\cdot 6 + \frac{1}{3}\cdot 1.50 \;=\; 4.50.
\end{aligned}$$

The estimate of the estimate is an average of averages, and it is not surprising that we can get the same result by a more comprehensive averaging. This more comprehensive averaging occurs when we estimate S_3 directly based on the information at time one:

$$\begin{aligned}
E_1[S_3](H) &= \frac{4}{9}\cdot 32 + \frac{2}{9}\cdot 8 + \frac{2}{9}\cdot 8 + \frac{1}{9}\cdot 2 = 18,\\
E_1[S_3](T) &= \frac{4}{9}\cdot 8 + \frac{2}{9}\cdot 2 + \frac{2}{9}\cdot 2 + \frac{1}{9}\cdot \frac{1}{2} = 4.50.
\end{aligned}$$

In conclusion, regardless of the outcome of the first toss, we have

$$\mathbb{E}_1\Big[\mathbb{E}_2[S_3]\Big] = \mathbb{E}_1[S_3].$$

Example 2.3.6 (Independence). The quotient $\frac{S_2}{S_1}$ takes either the value 2 or $\frac{1}{2}$, depending on whether the second coin toss results in head or tail, respectively. In particular, $\frac{S_2}{S_1}$ does not depend on the first coin toss. We compute

$$\begin{aligned}
\mathbb{E}_1\left[\frac{S_2}{S_1}\right](H) &= \frac{2}{3}\cdot \frac{S_2(HH)}{S_1(H)} + \frac{1}{3}\cdot \frac{S_2(HT)}{S_1(H)}\\
&= \frac{2}{3}\cdot 2 + \frac{1}{3}\cdot \frac{1}{2} \;=\; \frac{3}{2},\\
\mathbb{E}_1\left[\frac{S_2}{S_1}\right](T) &= \frac{2}{3}\cdot \frac{S_2(TH)}{S_1(T)} + \frac{1}{3}\cdot \frac{S_2(TT)}{S_1(T)}\\
&= \frac{2}{3}\cdot 2 + \frac{1}{3}\cdot \frac{1}{2} \;=\; \frac{3}{2}.
\end{aligned}$$

We see that $\mathbb{E}_1\left[\frac{S_2}{S_1}\right]$ does not depend on the first coin toss (is not really random) and in fact is equal to

$$\mathbb{E}\frac{S_2}{S_1} = \frac{2}{3}\cdot 2 + \frac{1}{3}\cdot\frac{1}{2} = \frac{3}{2}.$$

2.4 Martingales

In the binomial pricing model of Chapter 1, we chose risk-neutral probabilities $\tilde{p}$ and $\tilde{q}$ so that at every time n and for every coin toss sequence $\omega_1 \ldots \omega_n$ we have (2.3.3). In terms of the notation for conditional expectations introduced in Section 2.3, this fact can be written as (2.3.5). If we divide both sides of (2.3.5) by $(1+r)^n$, we get the equation

$$\frac{S_n}{(1+r)^n} = \widetilde{\mathbb{E}}_n\left[\frac{S_{n+1}}{(1+r)^{n+1}}\right]. \tag{2.4.1}$$

It does not matter in this model whether we write the term $\frac{1}{(1+r)^{n+1}}$ inside or outside the conditional expectation because it is constant (see Theorem 2.3.2(i)). In models with random interest rates, it would matter; we shall follow the practice of writing this term inside the conditional expectation since that is the way it would be written in models with random interest rates.

Equation (2.4.1) expresses the key fact that under the risk-neutral measure, for a stock that pays no dividend, the *best estimate based on the information at time n of the value of the discounted stock price at time $n+1$ is the discounted stock price at time n.* The risk-neutral probabilities are chosen to enforce this fact. Processes that satisfy this condition are called *martingales.* We give a formal definition of martingale under the actual probabilities p and q; the definition of martingale under the risk-neutral probabilities $\tilde{p}$ and $\tilde{q}$ is obtained by replacing $\mathbb{E}_n$ by $\widetilde{\mathbb{E}}_n$ in (2.4.2).

Definition 2.4.1. *Consider the binomial asset-pricing model. Let $M_0, M_1, \ldots, M_N$ be a sequence of random variables, with each M_n depending only on the first n coin tosses (and M_0 constant). Such a sequence of random variables is called an* adapted stochastic process.

(i) If

$$M_n = \mathbb{E}_n[M_{n+1}],\ n = 0, 1, \ldots, N-1, \tag{2.4.2}$$

we say this process is a martingale.

(ii) If

$$M_n \le \mathbb{E}_n[M_{n+1}],\ n = 0, 1, \ldots, N-1,$$

we say the process is a submartingale *(even though it may have a tendency to increase);*

(iii) If

$$M_n \geq \mathbb{E}_n[M_{n+1}],\ n = 0, 1, \ldots, N-1,$$

we say the process is a supermartingale *(even though it may have a tendency to decrease).*

Remark 2.4.2. The martingale property in (2.4.2) is a "one-step-ahead" condition. However, it implies a similar condition for any number of steps. Indeed, if $M_0, M_1, \ldots, M_N$ is a martingale and $n \leq N-2$, then the martingale property (2.4.2) implies

$$M_{n+1} = \mathbb{E}_{n+1}[M_{n+2}].$$

Taking conditional expectations on both sides based on the information at time n and using the iterated conditioning property (iii) of Theorem 2.3.2, we obtain

$$\mathbb{E}_n[M_{n+1}] = \mathbb{E}_n\Big[\mathbb{E}_{n+1}[M_{n+2}]\Big] = E_n[M_{n+2}].$$

Because of the martingale property (2.4.2), the left-hand side is M_n, and we thus have the "two-step-ahead" property

$$M_n = \mathbb{E}_n[M_{n+2}].$$

Iterating this argument, we can show that whenever $0 \leq n \leq m \leq N$,

$$M_n = \mathbb{E}_n[M_m]. \tag{2.4.3}$$

One might call this the "multistep-ahead" version of the martingale property.

Remark 2.4.3. The expectation of a martingale is constant over time, i.e., if $M_0, M_1, \ldots, M_N$ is a martingale, then

$$M_0 = \mathbb{E}M_n,\ n = 0, 1, \ldots, N. \tag{2.4.4}$$

Indeed, if $M_0, M_1, \ldots, M_N$ is a martingale, we may take expectations on both sides of (2.4.2), using Theorem 2.3.2(iii), and obtain $\mathbb{E}M_n = \mathbb{E}[M_{n+1}]$ for every n. It follows that

$$\mathbb{E}M_0 = \mathbb{E}M_1 = \mathbb{E}M_2 = \cdots = \mathbb{E}M_{N-1} = \mathbb{E}M_N.$$

But M_0 is not random, so $M_0 = \mathbb{E}M_0$, and (2.4.4) follows. □

In order to have a martingale, the equality in (2.4.2) must hold for all possible coin toss sequences. The stock price process in Figure 2.3.1 would be a martingale if the probability of an up move were $\hat{p} = \frac{1}{3}$ and the probability of a down move were $\hat{q} = \frac{2}{3}$ because, at every node in the tree in Figure 2.3.1, the stock price shown would then be the average of the two possible subsequent stock prices averaged with these weights. For example,

$$S_1(T) = 2 = \frac{1}{3} \cdot S_2(TH) + \frac{2}{3} \cdot S_2(TT).$$

A similar equation would hold at all other nodes in the tree, and therefore we would have a martingale under these probabilities.

A martingale has no tendency to rise or fall since the average of its next period values is always its value at the current time. Stock prices have a tendency to rise and, indeed, should rise on average faster than the money market in order to compensate investors for their inherent risk. In Figure 2.3.1 more realistic choices for p and q are $p = \frac{2}{3}$ and $q = \frac{1}{3}$. With these choices, we have

$$\mathbb{E}_n[S_{n+1}] = \frac{3}{2}S_n$$

at every node in the tree (i.e., on average, the next period stock price is 50% higher than the current stock price). This growth rate exceeds the 25% interest rate we have been using in this model, as it should. In particular, with $p = \frac{2}{3}$, $q = \frac{1}{3}$, and $r = \frac{1}{4}$, the discounted stock price has a tendency to rise. Note that when $r = \frac{1}{4}$, we have $\frac{1}{1+r} = \frac{4}{5}$, so the discounted stock price at time n is $\left(\frac{4}{5}\right)^n S_n$. We compute

$$\mathbb{E}_n\left[\left(\frac{4}{5}\right)^{n+1} S_{n+1}\right] = \left(\frac{4}{5}\right)^{n+1} \mathbb{E}_n[S_{n+1}] = \left(\frac{4}{5}\right)^n \cdot \frac{4}{5} \cdot \frac{3}{2} \cdot S_n \geq \left(\frac{4}{5}\right)^n S_n.$$

The discounted stock price is a submartingale under the actual probabilities $p = \frac{2}{3}$, $q = \frac{1}{3}$. This is typically the case in real markets.

The risk-neutral probabilities, on the other hand, are chosen to make the discounted stock price be a martingale. In Figure 2.3.1 with $\tilde{p} = \tilde{q} = \frac{1}{2}$, one can check that the martingale equation

$$\widetilde{\mathbb{E}}_n\left[\left(\frac{4}{5}\right)^{n+1} S_{n+1}\right] = \left(\frac{4}{5}\right)^n S_n \tag{2.4.5}$$

holds at every node. The following theorem shows that this example is representative.

Theorem 2.4.4. *Consider the general binomial model with* $0 < d < 1+r < u$. *Let the risk-neutral probabilities be given by*

$$\tilde{p} = \frac{1+r-d}{u-d}, \quad \tilde{q} = \frac{u-1-r}{u-d}.$$

Then, under the risk-neutral measure, the discounted stock price is a martingale, i.e., equation (2.4.1) holds at every time n and for every sequence of coin tosses.

We give two proofs of this theorem, an elementary one, which does not rely on Theorem 2.3.2, and a deeper one, which does rely on Theorem 2.3.2. The second proof will later be adapted to continuous-time models.

Note in Theorem 2.4.4 that the stock does not pay a dividend. For a dividend-paying stock, the situation is described in Exercise 2.10.

FIRST PROOF: Let n and $\omega_1 \dots \omega_n$ be given. Then

$$\begin{aligned}
&\widetilde{\mathbb{E}}_n\left[\frac{S_{n+1}}{(1+r)^{n+1}}\right](\omega_1\dots\omega_n)\\
&= \frac{1}{(1+r)^n}\cdot\frac{1}{1+r}\left[\tilde{p}S_{n+1}(\omega_1\dots\omega_n H)+\tilde{q}S_{n+1}(\omega_1\dots\omega_n T)\right]\\
&= \frac{1}{(1+r)^n}\cdot\frac{1}{1+r}\left[\tilde{p}uS_n(\omega_1\dots\omega_n)+\tilde{q}dS_n(\omega_1\dots\omega_n)\right]\\
&= \frac{S_n(\omega_1\dots\omega_n)}{(1+r)^n}\cdot\frac{\tilde{p}u+\tilde{q}d}{1+r}\\
&= \frac{S_n(\omega_1\dots\omega_n)}{(1+r)^n}.
\end{aligned}$$

SECOND PROOF: Note that $\frac{S_{n+1}}{S_n}$ depends only on the $(n+1)$st coin toss. Using the indicated properties from Theorem 2.3.2, we compute

$$\begin{aligned}
\widetilde{\mathbb{E}}_n\left[\frac{S_{n+1}}{(1+r)^{n+1}}\right] &= \widetilde{\mathbb{E}}_n\left[\frac{S_n}{(1+r)^{n+1}}\cdot\frac{S_{n+1}}{S_n}\right]\\
&= \frac{S_n}{(1+r)^n}\widetilde{\mathbb{E}}_n\left[\frac{1}{1+r}\cdot\frac{S_{n+1}}{S_n}\right]\\
&\qquad\text{(Taking out what is known)}\\
&= \frac{S_n}{(1+r)^n}\cdot\frac{1}{1+r}\widetilde{\mathbb{E}}\frac{S_{n+1}}{S_n}\\
&\qquad\text{(Independence)}\\
&= \frac{S_n}{(1+r)^n}\frac{\tilde{p}u+\tilde{q}d}{1+r}\\
&= \frac{S_n}{(1+r)^n}.
\end{aligned}$$

□

In a binomial model with N coin tosses, we imagine an investor who at each time n takes a position of Δ_n shares of stock and holds this position until time $n+1$, when he takes a new position of Δ_{n+1} shares. The portfolio rebalancing at each step is financed by investing or borrowing, as necessary, from the money market. The "portfolio variable" Δ_n may depend on the first n coin tosses, and Δ_{n+1} may depend on the first $n+1$ coin tosses. In other words, the *portfolio process* $\Delta_0, \Delta_1, \dots, \Delta_{N-1}$ is *adapted*, in the sense of Definition 2.4.1. If the investor begins with initial wealth X_0, and X_n denotes his wealth at each time n, then the evolution of his wealth is governed by the *wealth equation* (1.2.14) of Chapter 1, which we repeat here:

$$X_{n+1} = \Delta_n S_{n+1} + (1+r)(X_n - \Delta_n S_n),\ n = 0, 1, \dots, N-1. \qquad (2.4.6)$$

Note that each X_n depends only on the first n coin tosses (i.e., the *wealth process* is adapted).

We may inquire about the average rate of growth of the investor's wealth. If we mean the average under the actual probabilities, the answer depends on the portfolio process he uses. In particular, since a stock generally has a higher average rate of growth than the money market, the investor can achieve a rate of growth higher than the interest rate by taking long positions in the stock. Indeed, by borrowing from the money market, the investor can achieve an arbitrarily high *average* rate of growth. Of course, such leveraged positions are also extremely risky.

On the other hand, if we want to know the average rate of growth of the investor's wealth under the risk-neutral probabilities, the portfolio the investor uses is irrelevant. Under the risk-neutral probabilities, the average rate of growth of the stock is equal to the interest rate. No matter how the investor divides his wealth between the stock and the money market account, he will achieve an average rate of growth equal to the interest rate. Although some portfolio processes are riskier then others under the risk-neutral measure, they all have the same average rate of growth. We state this result as a theorem, whose proof is given in a way that we can later generalize to continuous time.

Theorem 2.4.5. *Consider the binomial model with N periods. Let $\Delta_0, \Delta_1, \dots, \Delta_{N-1}$ be an adapted portfolio process, let X_0 be a real number, and let the wealth process $X_1, \dots, X_N$ be generated recursively by (2.4.6). Then the discounted wealth process $\frac{X_n}{(1+r)^n}$, $n = 0, 1, \dots, N$, is a martingale under the risk-neutral measure; i.e.,*

$$\frac{X_n}{(1+r)^n} = \widetilde{\mathbb{E}}_n\left[\frac{X_{n+1}}{(1+r)^{n+1}}\right], \quad n = 0, 1, \dots, N-1. \tag{2.4.7}$$

PROOF: We compute

$$\begin{aligned}
\widetilde{\mathbb{E}}_n\left[\frac{X_{n+1}}{(1+r)^{n+1}}\right] &= \widetilde{\mathbb{E}}_n\left[\frac{\Delta_n S_{n+1}}{(1+r)^{n+1}} + \frac{X_n - \Delta_n S_n}{(1+r)^n}\right] \\
&= \widetilde{\mathbb{E}}_n\left[\frac{\Delta_n S_{n+1}}{(1+r)^{n+1}}\right] + \widetilde{\mathbb{E}}_n\left[\frac{X_n - \Delta_n S_n}{(1+r)^n}\right] \\
&\quad \text{(Linearity)} \\
&= \Delta_n \widetilde{\mathbb{E}}_n\left[\frac{S_{n+1}}{(1+r)^{n+1}}\right] + \frac{X_n - \Delta_n S_n}{(1+r)^n} \\
&\quad \text{(Taking out what is known)} \\
&= \Delta_n \frac{S_n}{(1+r)^n} + \frac{X_n - \Delta_n S_n}{(1+r)^n} \\
&\quad \text{(Theorem 2.4.4)} \\
&= \frac{X_n}{(1+r)^n}.
\end{aligned}$$

□

Corollary 2.4.6. *Under the conditions of Theorem 2.4.5, we have*

$$\widetilde{\mathbb{E}}\frac{X_n}{(1+r)^n} = X_0,\ n = 0,1,\ldots,N. \tag{2.4.8}$$

PROOF: The corollary follows from the fact that the expected value of a martingale cannot change with time and so must always be equal to the time-zero value of the martingale (see Remark 2.4.3). Applying this fact to the $\widetilde{\mathbb{P}}$-martingale $\frac{X_n}{(1+r)^n}$, $n = 0,1,\ldots,N$, we obtain (2.4.8). □

Theorem 2.4.5 and its corollary have two important consequences. The first is that there can be no arbitrage in the binomial model. If there were an arbitrage, we could begin with $X_0 = 0$ and find a portfolio process whose corresponding wealth process $X_1, X_2, \ldots, X_N$ satisfied $X_N(\omega) \geq 0$ for all coin toss sequences ω and $X_N(\overline{\omega}) > 0$ for at least one coin toss sequence $\overline{\omega}$. But then we would have $\widetilde{\mathbb{E}}X_0 = 0$ and $\widetilde{\mathbb{E}}\frac{X_N}{(1+r)^N} > 0$, which violates Corollary 2.4.6.

In general, if we can find a risk-neutral measure in a model (i.e., a measure that agrees with the actual probability measure about which price paths have zero probability, and under which the discounted prices of all primary assets are martingales), then there is no arbitrage in the model. This is sometimes called the *First Fundamental Theorem of Asset Pricing.* The essence of its proof is contained in the preceding paragraph: under a risk-neutral measure, the discounted wealth process has constant expectation, so it cannot begin at zero and later be strictly positive with positive probability unless it also has a positive probability of being strictly negative. The First Fundamental Theorem of Asset Pricing will prove useful for ruling out arbitrage in term-structure models later on and thereby lead to the Heath-Jarrow-Morton no-arbitrage condition on forward rates.

The other consequence of Theorem 2.4.5 is the following version of the *risk-neutral pricing formula.* Let V_N be a random variable (derivative security paying off at time N) depending on the first N coin tosses. We know from Theorem 1.2.2 of Chapter 1 that there is an initial wealth X_0 and a replicating portfolio process $\Delta_0, \ldots, \Delta_{N-1}$ that generates a wealth process $X_1, \ldots, X_N$ satisfying $X_N = V_N$, no matter how the coin tossing turns out. Because $\frac{X_n}{(1+r)^n}$, $n = 0,1,\ldots,N$, is a martingale, the "multistep ahead" property of Remark 2.4.2 implies

$$\frac{X_n}{(1+r)^n} = \widetilde{\mathbb{E}}_n\left[\frac{X_N}{(1+r)^N}\right] = \widetilde{\mathbb{E}}_n\left[\frac{V_N}{(1+r)^N}\right]. \tag{2.4.9}$$

According to Definition 1.2.3 of Chapter 1, we define the price of the derivative security at time n to be X_n and denote this price by the symbol V_n. Thus, (2.4.9) may be rewritten as

$$\frac{V_n}{(1+r)^n} = \widetilde{\mathbb{E}}_n\left[\frac{V_N}{(1+r)^N}\right] \tag{2.4.10}$$

or, equivalently,

$$V_n = \widetilde{\mathbb{E}}_n\left[\frac{V_N}{(1+r)^{N-n}}\right]. \tag{2.4.11}$$

We summarize with a theorem.

Theorem 2.4.7 (Risk-neutral pricing formula). *Consider an N-period binomial asset-pricing model with $0 < d < 1 + r < u$ and with risk-neutral probability measure $\widetilde{\mathbb{P}}$. Let V_N be a random variable (a derivative security paying off at time N) depending on the coin tosses. Then, for n between 0 and N, the price of the derivative security at time n is given by the* risk-neutral pricing formula *(2.4.11). Furthermore, the discounted price of the derivative security is a martingale under $\widetilde{\mathbb{P}}$; i.e.,*

$$\frac{V_n}{(1+r)^n} = \widetilde{\mathbb{E}}_n\left[\frac{V_{n+1}}{(1+r)^{n+1}}\right], \quad n = 0, 1, \ldots, N-1. \tag{2.4.12}$$

The random variables V_n defined by (2.4.11) are the same as the random variable V_n defined in Theorem 1.2.2.

The remaining steps in the proof of Theorem 2.4.7 are outlined in Exercise 2.8. We note that we chose the risk-neutral measure in order to make the discounted stock price a martingale. According to Theorem 2.4.7, a consequence of this is that discounted derivative security prices under the risk-neutral measure are also martingales.

So far, we have discussed only derivative securities that pay off on a single date. Many securities, such as coupon-paying bonds and interest rate swaps, make a series of payments. For such a security, we have the following pricing and hedging formulas.

Theorem 2.4.8 (Cash flow valuation). *Consider an N-period binomial asset pricing-model with $0 < d < 1 + r < u$, and with risk-neutral probability measure $\widetilde{\mathbb{P}}$. Let $C_0, C_1, \ldots, C_N$ be a sequence of random variables such that each C_n depends only on $\omega_1 \ldots \omega_n$. The price at time n of the derivative security that makes payments $C_n, \ldots, C_N$ at times $n, \ldots, N$, respectively, is*

$$V_n = \widetilde{\mathbb{E}}_n\left[\sum_{k=n}^{N} \frac{C_k}{(1+r)^{k-n}}\right], \quad n = 0, 1, \ldots, N. \tag{2.4.13}$$

The price process V_n, $n = 0, 1, \ldots, N$, satisfies

$$C_n(\omega_1 \ldots \omega_n) = V_n(\omega_1 \ldots \omega_n) - \frac{1}{1+r}\Big[\tilde{p}V_{n+1}(\omega_1 \ldots \omega_n H) + \tilde{q}V_{n+1}(\omega_1 \ldots \omega_n T)\Big]. \tag{2.4.14}$$

We define

$$\Delta_n(\omega_1 \ldots \omega_n) = \frac{V_{n+1}(\omega_1 \ldots \omega_n H) - V_{n+1}(\omega_1 \ldots \omega_n T)}{S_{n+1}(\omega_1 \ldots \omega_n H) - S_{n+1}(\omega_1 \ldots \omega_n T)}, \tag{2.4.15}$$

where n ranges between 0 and $N-1$. If we set $X_0 = V_0$ and define recursively forward in time the portfolio values $X_1, X_2, \ldots, X_N$ by

$$X_{n+1} = \Delta_n S_{n+1} + (1+r)(X_n - C_n - \Delta_n S_n), \tag{2.4.16}$$

then we have

$$X_n(\omega_1 \ldots \omega_n) = V_n(\omega_1 \ldots \omega_n) \tag{2.4.17}$$

for all n and all $\omega_1 \ldots \omega_n$.

In Theorem 2.4.8, V_n is the so-called *net present value* at time n of the sequence of payments $C_n, \ldots, C_N$. It is just the sum of the value $\widetilde{\mathbb{E}}_n\left[\frac{C_k}{(1+r)^{(n-k)}}\right]$ of each of the payments C_k to be made at times $k = n, k = n+1, \ldots, k = N$. Note that the payment at time n is included. This payment C_n depends on only the first n tosses and so can be taken outside the conditional expectation $\widetilde{\mathbb{E}}_n$, i.e.,

$$V_n = C_n + \widetilde{\mathbb{E}}_n\left[\sum_{k=n+1}^{N} \frac{C_k}{(1+r)^{k-n}}\right], n = 0, 1, \ldots, N-1. \tag{2.4.18}$$

In the case of $n = N$, (2.4.13) reduces to

$$V_N = C_N. \tag{2.4.19}$$

Consider an agent who is short the cash flows represented by $C_0, \ldots, C_n$ (i.e., an agent who must make the payment C_n at each time n). (We allow these payments to be negative as well as positive. If a payment is negative, the agent who is short actually receives cash.) Suppose the agent in the short position invests in the stock and money market account, so that, at time n, before making the payment C_n, the value of his portfolio is X_n. He then makes the payment C_n. Suppose he then takes a position Δ_n in stock. This will cause the value of his portfolio at time $n+1$ before making the payment C_{n+1} to be X_{n+1}, given by (2.4.16). If this agent begins with $X_0 = V_0$ and chooses his stock positions Δ_n by (2.4.15), then (2.4.17) holds and, in particular, $X_N = V_N = C_N$ (see (2.4.17), and (2.4.19)). Then, at time N he makes the final payment C_N and is left with 0. He has perfectly hedged the short position in the cash flows. This is the justification for calling V_n the value at time n of the future cash flows, including the payment C_n to be made at time n.

PROOF OF THEOREM 2.4.8: To prove (2.4.17), we proceed by induction on n. The induction hypothesis is that $X_n(\omega_1 \ldots \omega_n) = V_n(\omega_1 \ldots \omega_n)$ for some $n \in \{0, 1, \ldots, N-1\}$ and all $\omega_1 \ldots \omega_n$. We need to show that

$$X_{n+1}(\omega_1 \ldots \omega_n H) = V_{n+1}(\omega_1 \ldots \omega_n H), \tag{2.4.20}$$
$$X_{n+1}(\omega_1 \ldots \omega_n T) = V_{n+1}(\omega_1 \ldots \omega_n T). \tag{2.4.21}$$

We prove (2.4.20); the proof of (2.4.21) is analogous.

From (2.4.18) and iterated conditioning (Theorem 2.3.2(iii)), we have

$$\begin{aligned}V_n &= C_n + \widetilde{\mathbb{E}}_n\left[\frac{1}{1+r}\widetilde{\mathbb{E}}_{n+1}\left[\sum_{k=n+1}^{N}\frac{C_k}{(1+r)^{k-(n+1)}}\right]\right]\\ &= C_n + \widetilde{\mathbb{E}}_n\left[\frac{1}{1+r}V_{n+1}\right],\end{aligned}$$

where we have used (2.4.13) with n replaced by $n+1$ in the last step. In other words, for all $\omega_1 \ldots \omega_n$, we have

$$\begin{aligned}&V_n(\omega_1\ldots\omega_n) - C_n(\omega_1\ldots\omega_n)\\ &= \frac{1}{1+r}\Big[\tilde{p}V_{n+1}(\omega_1\ldots\omega_n H) + \tilde{q}V_{n+1}(\omega_1\ldots\omega_n T)\Big].\end{aligned}$$

Since $\omega_1 \ldots \omega_n$ will be fixed for the rest of the proof, we will suppress these symbols. For example, the last equation will be written simply as

$$V_n - C_n = \frac{1}{1+r}\Big[\tilde{p}V_{n+1}(H) + \tilde{q}V_{n+1}(T)\Big].$$

We compute

$$\begin{aligned}X_{n+1}(H) &= \Delta_n S_{n+1}(H) + (1+r)(X_n - C_n - \Delta_n S_n)\\ &= \frac{V_{n+1}(H) - V_{n+1}(T)}{S_{n+1}(H) - S_{n+1}(T)}\big(S_{n+1}(H) - (1+r)S_n\big)\\ &\qquad + (1+r)(V_n - C_n)\\ &= \frac{V_{n+1}(H) - V_{n+1}(T)}{(u-d)S_n}\big(uS_n - (1+r)S_n\big)\\ &\qquad + \tilde{p}V_{n+1}(H) + \tilde{q}V_{n+1}(T)\\ &= \big(V_{n+1}(H) - V_{n+1}(T)\big)\frac{u-1-r}{u-d} + \tilde{p}V_{n+1}(H) + \tilde{q}V_{n+1}(T)\\ &= \big(V_{n+1}(H) - V_{n+1}(T)\big)\tilde{q} + \tilde{p}V_{n+1}(H) + \tilde{q}V_{n+1}(T)\\ &= (\tilde{p}+\tilde{q})V_{n+1}(H) \;=\; V_{n+1}(H).\end{aligned}$$

This is (2.4.20). □

2.5 Markov Processes

In Section 1.3, we saw that the computational requirements of the derivative security pricing algorithm of Theorem 1.2.2 can often be substantially reduced by thinking carefully about what information needs to be remembered as we go from period to period. In Example 1.3.1 of Section 1.3, the stock price was relevant, but the path it followed to get to its current price was not. In Example 1.3.2 of Section 1.3, the stock price and the maximum value it had achieved up to the current time were relevant. In this section, we formalize the procedure for determining what is relevant and what is not.

Definition 2.5.1. *Consider the binomial asset-pricing model. Let $X_0, X_1, \ldots, X_N$ be an adapted process. If, for every n between 0 and $N-1$ and for every function $f(x)$, there is another function $g(x)$ (depending on n and f) such that*

$$\mathbb{E}_n[f(X_{n+1})] = g(X_n), \tag{2.5.1}$$

we say that $X_0, X_1, \ldots, X_N$ is a Markov process.

By definition, $\mathbb{E}_n[f(X_{n+1})]$ is random; it depends on the first n coin tosses. The Markov property says that this dependence on the coin tosses occurs through X_n (i.e., the information about the coin tosses one needs in order to evaluate $\mathbb{E}_n[f(X_{n+1})]$ is summarized by X_n). We are not so concerned with determining a formula for the function g right now as we are with asserting its existence because its mere existence tells us that if the payoff of a derivative security is random only through its dependence on X_N, then there is a version of the derivative security pricing algorithm in which we do not need to store path information (see Theorem 2.5.8). In examples in this section, we shall develop a method for finding the function g.

Example 2.5.2 (Stock price). In the binomial model, the stock price at time $n+1$ is given in terms of the stock price at time n by the formula

$$S_{n+1}(\omega_1 \ldots \omega_n \omega_{n+1}) = \begin{cases} uS_n(\omega_1 \ldots \omega_n), \text{ if } \omega_{n+1} = H, \\ dS_n(\omega_1 \ldots \omega_n), \text{ if } \omega_{n+1} = T. \end{cases}$$

Therefore,

$$\mathbb{E}_n[f(S_{n+1})](\omega_1 \ldots \omega_n) = pf(uS_n(\omega_1 \ldots \omega_n)) + qf(dS_n(\omega_1 \ldots \omega_n)),$$

and the right-hand side depends on $\omega_1 \ldots \omega_n$ only through the value of $S_n(\omega_1 \ldots \omega_n)$. Omitting the coin tosses $\omega_1 \ldots \omega_n$, we can rewrite this equation as

$$\mathbb{E}_n[f(S_{n+1})] = g(S_n),$$

where the function $g(x)$ of the dummy variable x is defined by $g(x) = pf(ux) + qf(dx)$. This shows that the stock price process is Markov.

Indeed, the stock price process is Markov under either the actual or the risk-neutral probability measure. To determine the price V_n at time n of a derivative security whose payoff at time N is a function v_N of the stock price S_N (i.e., $V_N = v_N(S_N)$), we use the risk-neutral pricing formula (2.4.12), which reduces to

$$V_n = \frac{1}{1+r}\widetilde{\mathbb{E}}_n[V_{n+1}], \; n = 0, 1, \ldots, N-1.$$

But $V_N = v_N(S_N)$ and the stock price process is Markov, so

$$V_{N-1} = \frac{1}{1+r}\widetilde{\mathbb{E}}_{N-1}[v_N(S_N)] = v_{N-1}(S_{N-1})$$

for some function v_{N-1}. Similarly,

$$V_{N-2} = \frac{1}{1+r}\widetilde{\mathbb{E}}_{N-2}[v_{N-1}(S_{N-1})] = v_{N-2}(S_{N-2})$$

for some function v_{N-2}. In general, $V_n = v_n(S_n)$ for some function v_n. Moreover, we can compute these functions recursively by the algorithm

$$v_n(s) = \frac{1}{1+r}\Big[\tilde{p}v_{n+1}(us) + \tilde{q}v_{n+1}(ds)\Big], \quad n = N-1, N-2, \ldots, 0. \tag{2.5.2}$$

This algorithm works in the binomial model for any derivative security whose payoff at time N is a function only of the stock price at time N. In particular, we have the same algorithm for puts and calls. The only difference is in the formula for $v_N(s)$. For the call, we have $v_N(s) = (s-K)^+$; for the put, we have $v_N(s) = (K-s)^+$. □

The martingale property is the special case of (2.5.1) with $f(x) = x$ and $g(x) = x$. In order for a process to be Markov, it is necessary that for *every* function f there must be a corresponding function g such that (2.5.1) holds. Not every martingale is Markov. On the other hand, even when considering the function $f(x) = x$, the Markov property requires only that $\mathbb{E}_n[M_{n+1}] = g(M_n)$ for some function g; it does not require that the function g be given by $g(x) = x$. Not every Markov process is a martingale. Indeed, Example 2.5.2 shows that the stock price is Markov under both the actual and the risk-neutral probability measures. It is typically not a martingale under either of these measures. However, if $pu + qd = 1$, then the stock price is both a martingale and a Markov process under the actual probability measure.

The following lemma often provides the key step in the verification that a process is Markov.

Lemma 2.5.3 (Independence). *In the N-period binomial asset pricing model, let n be an integer between 0 and N. Suppose the random variables $X^1, \ldots, X^K$ depend only on coin tosses 1 through n and the random variables $Y^1, \ldots, Y^L$ depend only on coin tosses $n+1$ through N. (The superscripts $1, \ldots, K$ on X and $1, \ldots, L$ on Y are superscripts, not exponents.) Let $f(x^1, \ldots, x^K, y^1, \ldots, y^L)$ be a function of dummy variables $x^1, \ldots, x^K$ and $y^1, \ldots, y^L$, and define*

$$g(x^1, \ldots, x^K) = \mathbb{E}f(x^1, \ldots, x^K, Y^1, \ldots, Y^L). \tag{2.5.3}$$

Then

$$\mathbb{E}_n[f(X^1, \ldots, X^K, Y^1, \ldots, Y^L)] = g(X^1, \ldots, X^K). \tag{2.5.4}$$

For the following discussion and proof of the lemma, we assume that $K = L = 1$. Then (2.5.3) takes the form

$$g(x) = \mathbb{E}f(x, Y) \tag{2.5.3$'$}$$

and (2.5.4) takes the form

$$\mathbb{E}_n[f(X,Y)] = g(X), \tag{2.5.4$'$}$$

where the random variable X is assumed to depend only on the first n coin tosses, and the random variable Y depends only on coin tosses $n+1$ through N.

This lemma generalizes the property "taking out what is known" of Theorem 2.3.2(ii). Since X is "known" at time n, we want to "take it out" of the computation of the conditional expectation $\mathbb{E}_n[f(X,Y)]$. However, because X is inside the argument of the function f, we cannot simply factor it out as we did in Theorem 2.3.2(ii). Therefore, we hold it constant by replacing the random variable X by an arbitrary but fixed dummy variable x. We then compute the conditional expectation of the random variable $f(x,Y)$, whose randomness is due only to the dependence of Y on tosses $n+1$ through N. Because of Theorem 2.3.2(iv), this conditional expectation is the same as the unconditional expectation in (2.5.3)$'$. Finally, we recall that $E_n[f(X,Y)]$ must depend on the value of the random variable X, so we replace the dummy variable x by the random variable X after g is computed.

In the context of Example 2.5.2, we can take $X = S_n$, which depends only on the first n coin tosses, and take $Y = \frac{S_{n+1}}{S_n}$, which depends only on the $(n+1)$st coin toss, taking the value u if the $(n+1)$st toss results in a head and taking the value d if it results in a tail. We are asked to compute

$$\mathbb{E}_n[f(S_{n+1})] = \mathbb{E}_n[f(XY)].$$

We replace X by a dummy variable x and compute

$$g(x) = \mathbb{E}f(xY) = pf(ux) + qf(dx).$$

Then $\mathbb{E}_n[f(S_{n+1}] = g(S_n)$.

PROOF OF LEMMA 2.5.3: Let $\omega_1 \dots \omega_n$ be fixed but arbitrary. By the definition (2.3.6) of conditional expectation,

$$\begin{aligned}
&\mathbb{E}_n[f(X,Y)](\omega_1 \dots \omega_n)\\
&= \sum_{\omega_{n+1}\dots\omega_N} f(X(\omega_1 \dots \omega_n), Y(\omega_{n+1} \dots \omega_N))p^{\#H(\omega_{n+1}\dots\omega_N)}q^{\#T(\omega_{n+1}\dots\omega_N)},
\end{aligned}$$

whereas

$$\begin{aligned}
g(x) &= \mathbb{E}f(x,Y)\\
&= \sum_{\omega_{n+1}\dots\omega_N} f(x, Y(\omega_{n+1} \dots \omega_N))p^{\#H(\omega_{n+1}\dots\omega_N)}q^{\#T(\omega_{n+1}\dots\omega_N)}.
\end{aligned}$$

It is apparent that

$$\mathbb{E}_n[f(X,Y)](\omega_1 \dots \omega_N) = g(X(\omega_1 \dots \omega_N)).$$

□

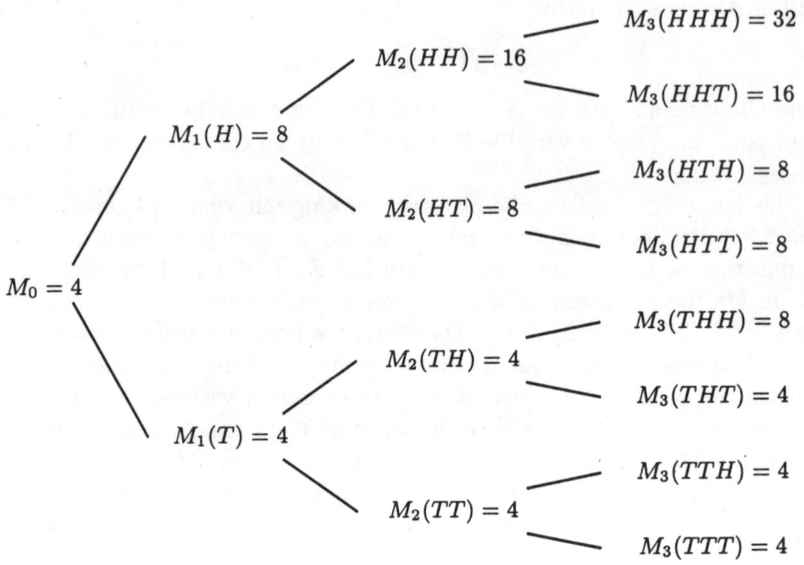

Fig. 2.5.1. The maximum stock price to date.

Example 2.5.4 (Non-Markov process). In the binomial model of Figure 2.3.1, consider the maximum-to-date process $M_n = \max_{0 \le k \le n} S_k$, shown in Figure 2.5.1. With $p = \frac{2}{3}$ and $q = \frac{1}{3}$, we have

$$\mathbb{E}_2[M_3](TH) = \frac{2}{3}M_3(THH) + \frac{1}{3}M_3(THT) = \frac{16}{3} + \frac{4}{3} = 6\frac{2}{3},$$

but

$$\mathbb{E}_2[M_3](TT) = \frac{2}{3}M_3(TTH) + \frac{1}{3}M_3(TTT) = \frac{8}{3} + \frac{4}{3} = 4.$$

Since $M_2(TH) = M_2(TT) = 4$, there cannot be a function g such that $\mathbb{E}_2[M_3](TH) = g(M_2(TH))$ and $\mathbb{E}_2[M_3](TT) = g(M_2(TT))$. The right-hand sides would be the same, but the left-hand sides would not. The maximum-to-date process is not Markov because recording only that the value of the maximum-to-date at time two is 4, without recording the value of the stock price at time two, neglects information relevant to the evolution of the maximum-to-date process after time two. □

When we encounter a non-Markov process, we can sometimes recover the Markov property by adding one or more so-called *state variables*. The term "state variable" is used because if we can succeed in recovering the Markov property by adding these variables, we will have determined a way to describe the "state" of the market in terms of these variables. This approach to recovering the Markov property requires that we generalize Definition 2.5.1 to multidimensional processes.

Definition 2.5.5. *Consider the binomial asset-pricing model. Let* $\{(X_n^1, \ldots, X_n^K); n = 0, 1, \ldots, N\}$ *be a* K*-dimensional adapted process; i.e.,* K *one-dimensional adapted processes. If, for every* n *between* 0 *and* $N-1$ *and for every function* $f(x^1, \ldots, x^K)$*, there is another function* $g(x^1, \ldots, x^K)$ *(depending on* n *and* f*) such that*

$$\mathbb{E}_n[f(X_{n+1}^1, \ldots, X_{n+1}^K)] = g(X_n^1, \ldots, X_n^K), \tag{2.5.5}$$

we say that $\{(X_n^1, \ldots, X_n^K); n = 0, 1, \ldots, N\}$ *is a* K-dimensional Markov process.

Example 2.5.6. In an N-period binomial model, consider the two-dimensional adapted process $\{(S_n, M_n); n = 0, 1, \ldots, N\}$, where S_n is the stock price at time n and $M_n = \max_{0 \le k \le n} S_k$ is the stock price maximum-to-date. We show that this two-dimensional process is Markov. To do that, we define $Y = \frac{S_{n+1}}{S_n}$, which depends only on the $(n+1)$st coin toss. Then

$$S_{n+1} = S_n Y$$

and

$$M_{n+1} = M_n \vee S_{n+1} = M_n \vee (S_n Y),$$

where $x \vee y = \max\{x, y\}$. We wish to compute

$$\mathbb{E}_n[f(S_{n+1}, M_{n+1})] = \mathbb{E}_n[f(S_n Y, M_n \vee (S_n Y))].$$

According to Lemma 2.5.3, we replace S_n by a dummy variable s, replace M_n by a dummy variable m, and compute

$$g(s, m) = \mathbb{E} f(sY, m \vee (sY)) = pf(us, m \vee (us)) + qf(ds, m \vee (ds)).$$

Then

$$\mathbb{E}_n[f(S_{n+1}, M_{n+1})] = g(S_n, M_n).$$

Since we have obtained a formula for $\mathbb{E}_n[f(S_{n+1}, M_{n+1})]$ in which the only randomness enters through the random variables S_n and M_n, we conclude that the two-dimensional process is Markov. In this example, we have used the actual probability measure, but the same argument shows that $\{(S_n, M_n); n = 0, 1, \ldots, N\}$ is Markov under the risk-neutral probability measure $\widetilde{\mathbb{P}}$. □

Remark 2.5.7. The Markov property, in both the one-dimensional form of Definition 2.5.1 and the multidimensional form of Definition 2.5.5, is a "one-step-ahead" property, determining a formula for the conditional expectation of X_{n+1} in terms of X_n. However, it implies a similar condition for any number of steps. Indeed, if $X_0, X_1, \ldots, X_N$ is a Markov process and $n \le N-2$, then the "one-step-ahead" Markov property implies that for every function h there is a function f such that

$$\mathbb{E}_{n+1}[h(X_{n+2})] = f(X_{n+1}).$$

Taking conditional expectations on both sides based on the information at time n and using the iterated conditioning property (iii) of Theorem 2.3.3, we obtain

$$\mathbb{E}_n[h(X_{n+2})] = \mathbb{E}_n\Big[\mathbb{E}_{n+1}[h(X_{n+2})]\Big] = \mathbb{E}_n[f(X_{n+1})].$$

Because of the "one-step-ahead" Markov property, the right-hand side is $g(X_n)$ for some function g, and we have obtained the "two-step-ahead" Markov property

$$\mathbb{E}_n[h(X_{n+2})] = g(X_n).$$

Iterating this argument, we can show that whenever $0 \leq n \leq m \leq N$ and h is any function, then there is another function g such that the "multi-step-ahead" Markov property

$$\mathbb{E}_n[h(X_m)] = g(X_n) \tag{2.5.6}$$

holds. Similarly, if $\{(X_n^1, \ldots, X_n^K); n = 1, 2, \ldots, N\}$ is a K-dimensional Markov process, then whenever $0 \leq n \leq m \leq N$ and $h(x^1, \ldots, x^K)$ is any function, there is another function $g(x^1, \ldots, x^K)$ such that

$$\mathbb{E}_n[h(X_m^1, \ldots, X_m^K)] = g(X_n^1, \ldots, X_n^K). \tag{2.5.7}$$

□

In the binomial pricing model, suppose we have a Markov process $X_0, X_1, \ldots, X_N$ under the risk-neutral probability measure $\widetilde{\mathbb{P}}$, and we have a derivative security whose payoff V_N at time N is a function v_N of X_N, i.e., $V_N = v_N(X_N)$. The difference between V_N and v_N is that the argument of the former is $\omega_1 \ldots \omega_N$, a sequence of coin tosses, whereas the argument of the latter is a real number, which we will sometimes denote by the dummy variable x. In particular, there is nothing random about $v_N(x)$. However, if in place of the dummy variable x we substitute the random variable X_N (actually $X_N(\omega_1 \ldots \omega_N)$), then we have a random variable. Indeed, we have

$$V_N(\omega_1 \ldots \omega_N) = v_N(X_N(\omega_1 \ldots \omega_N)) \text{ for all } \omega_1 \ldots \omega_N.$$

The risk-neutral pricing formula (2.4.11) says that the price of this derivative security at earlier times n is

$$V_n(\omega_1 \ldots \omega_n) = \widetilde{\mathbb{E}}_n\left[\frac{V_N}{(1+r)^{N-n}}\right](\omega_1 \ldots \omega_n) \text{ for all } \omega_1 \ldots \omega_n.$$

On the other hand, the "multi-step-ahead" Markov property implies that there is a function v_n such that

$$\widetilde{\mathbb{E}}_n\left[\frac{V_N}{(1+r)^{N-n}}\right](\omega_1 \ldots \omega_n) = v_n\big(X_n(\omega_1 \ldots \omega_n)\big) \text{ for all } \omega_1 \ldots \omega_n.$$

Therefore, the price of the derivative security at time n is a function of X_n, i.e.,

$$V_n = v_n(X_n).$$

Instead of computing the random variables V_n, we can compute the functions v_n, and this is generally much more manageable computationally. In particular, when the Markov process $X_0, X_1, \dots, X_N$ is the stock price itself, we get the algorithm (2.5.2).

The same idea can be used for multidimensional Markov processes under $\widetilde{\mathbb{P}}$. A case of this was Example 1.3.2 of Section 1.3, in which the payoff of a derivative security was $V_3 = M_3 - S_3$, the difference between the stock price at time three and its maximum between times zero and three. Because only the stock price and its maximum-to-date appear in the payoff, we can use the two-dimensional Markov process $\{(S_n, M_n); n = 0, 1, 2, 3\}$ to treat this problem, which was done implicitly in that example.

Here we generalize Example 1.3.2 to an N-period binomial model with a derivative security whose payoff at time N is a function $v_N(S_N, M_N)$ of the stock price and the maximum stock price. (We do not mean that v_N is necessarily a function of *both* S_N and M_N but rather that these are the *only* random variables on which V_N depends. For example, we could have $V_N = (M_N - K)^+$. Even though the stock price does not appear in this particular V_N, we would need it to execute the pricing algorithm (2.5.9) below because the maximum-to-date process is not Markov by itself.) According to the "multi-step-ahead" Markov property, for any n between zero and N, there is a (nonrandom) function $v_n(s, m)$ such that the price of the option at time n is

$$V_n = v_n(S_n, M_n) = \widetilde{\mathbb{E}}_n \left[\frac{v_N(S_N, M_N)}{(1+r)^{N-n}} \right].$$

We can use the Independence Lemma 2.5.3 to derive an algorithm for computing the functions v_n. We always have the risk-neutral pricing formula (see (2.4.12))

$$V_n = \frac{1}{1+r} \widetilde{\mathbb{E}}_n [V_{n+1}]$$

relating the price of a derivative security at time n to its price at time $n+1$. Suppose that for some n between zero and $N-1$, we have computed the function v_{n+1} such that $V_{n+1} = v_{n+1}(S_{n+1}, M_{n+1})$. Then

$$\begin{aligned} V_n &= \frac{1}{1+r} \widetilde{\mathbb{E}}_n [V_{n+1}] \\ &= \frac{1}{1+r} \widetilde{\mathbb{E}}_n \left[v_{n+1}(S_{n+1}, M_{n+1}) \right] \\ &= \frac{1}{1+r} \widetilde{\mathbb{E}}_n \left[v_{n+1} \left(S_n \cdot \frac{S_{n+1}}{S_n}, M_n \vee \left(S_n \cdot \frac{S_{n+1}}{S_n} \right) \right) \right]. \end{aligned}$$

To compute this last expression, we replace S_n and M_n by dummy variables s and m because they depend only on the first n tosses. We then take the unconditional expectation of $\frac{S_{n+1}}{S_n}$ because it does not depend on the first n tosses, i.e., we define

$$v_n(s,m) = \frac{1}{1+r}\widetilde{\mathbb{E}}_n\left[v_{n+1}\left(s\cdot\frac{S_{n+1}}{S_n}, m \vee \left(s\cdot\frac{S_{n+1}}{S_n}\right)\right)\right] \tag{2.5.8}$$
$$= \frac{1}{1+r}\Big[\tilde{p}v_{n+1}(us, m\vee(us)) + \tilde{q}v_{n+1}(ds, m\vee(ds))\Big].$$

The Independence Lemma 2.5.3 asserts that $V_n = v_n(S_n, M_n)$.

We will only need to know the value of $v_n(s,m)$ when $m \geq s$ since $M_n \geq S_n$. We can impose this condition in (2.5.8). But when $m \geq s$, if $d \leq 1$ as it usually is, we have $m \vee (ds) = m$. Therefore, we can rewrite (2.5.8) as

$$v_n(s,m) = \frac{1}{1+r}\Big[\tilde{p}v_{n+1}(us, m\vee(us)) + \tilde{q}v_{n+1}(ds, m)\Big], \tag{2.5.9}$$
$$m \geq s > 0,\ n = N-1, N-2, \ldots, 0.$$

This algorithm works for any derivative security whose payoff at time N depends only on the random variables S_N and M_N.

In Example 1.3.2, we were given that $V_3 = v_3(s,m)$, where $v_3(s,m) = m - s$. We used (2.5.9) to compute v_2, then used it again to compute v_1, and finally used it to compute v_0. These steps were carried out in Example 1.3.2.

In continuous time, we shall see that the analogue of recursive equations (2.5.9) are partial differential equations. The process that gets us from the continuous-time analogue of the risk-neutral pricing formula to these partial differential equations is the *Feynman-Kac Theorem.*

We summarize this discussion with a theorem.

Theorem 2.5.8. *Let $X_0, X_1, \ldots, X_N$ be a Markov process under the risk-neutral probability measure $\widetilde{\mathbb{P}}$ in the binomial model. Let $v_N(x)$ be a function of the dummy variable x, and consider a derivative security whose payoff at time N is $v_N(X_N)$. Then, for each n between 0 and N, the price V_n of this derivative security is some function v_n of X_n, i.e.,*

$$V_n = v_n(X_n),\ n = 0, 1, \ldots, N. \tag{2.5.10}$$

There is a recursive algorithm for computing v_n whose exact formula depends on the underlying Markov process $X_0, X_1, \ldots, X_N$. Analogous results hold if the underlying Markov process is multidimensional.

2.6 Summary

This chapter sets out the view of probability that begins with a random experiment having outcome ω. The collection of all possible outcomes is called the *sample space* Ω, and on this space we have a probability measure $\mathbb{P}$. When Ω is finite, we describe $\mathbb{P}$ by specifying for each $\omega \in \Omega$ the probability $\mathbb{P}(\omega)$ assigned to ω by $\mathbb{P}$. A random variable is a function X from Ω to $\mathbb{R}$, and the expectation of the random variable X is $\mathbb{E}X = \sum_{\omega\in\Omega} X(\omega)\mathbb{P}(\omega)$. If we have a second probability measure $\widetilde{\mathbb{P}}$ on Ω, then we will have another way of computing the expectation, namely $\widetilde{\mathbb{E}}X = \sum_{\omega\in\Omega} X(\omega)\widetilde{\mathbb{P}}(\omega)$. The random variable X is the same in both cases, even though the two expectations are different. The point is that the random variable should not be thought of as a distribution. When we change probability measures, distributions (and hence expectations) will change, but random variables will not.

In the binomial model, we may see coin tosses $\omega_1 \dots \omega_n$ and, based on this information, compute the conditional expectation of a random variable X that depends on coin tosses $\omega_1 \dots \omega_n\omega_{n+1} \dots \omega_N$. This is done by averaging over the possible outcomes of the "remaining" coin tosses $\omega_{n+1} \dots \omega_N$. If we are computing the conditional expectation under the risk-neutral probabilities, this results in the formula

$$\widetilde{\mathbb{E}}_n[X](\omega_1 \dots \omega_n) \tag{2.3.6}$$

$$= \sum_{\omega_{n+1}\dots\omega_N} \tilde{p}^{\#H(\omega_{n+1}\dots\omega_N)}\tilde{q}^{\#T(\omega_{n+1}\dots\omega_N)} X(\omega_1 \dots \omega_n\omega_{n+1} \dots \omega_N).$$

This conditional expectation is a random variable because it depends on the first n coin tosses $\omega_1 \dots \omega_n$. Conditional expectations have five fundamental properties, which are provided in Theorem 2.3.2.

In a multiperiod binomial model, a *martingale* under the risk-neutral probability measure $\widetilde{\mathbb{P}}$ is a sequence of random variables $M_0, M_1, \dots, M_N$, where each M_n depends on only the first n coin tosses, and

$$M_n(\omega_1 \dots \omega_n) = \widetilde{\mathbb{E}}_n[M_{n+1}](\omega_1 \dots \omega_n)$$

no matter what the value of n and no matter what the coin tosses $\omega_1 \dots \omega_n$ are. A martingale has no tendency to rise or fall. Conditioned on the information we have at time n, the expected value of the martingale at time $n+1$ is its value at time n.

Under the risk-neutral probability measure, the discounted stock price is a martingale, as is the discounted value of any portfolio that trades in the stock and money markets account. In particular, if X_n is the value of a portfolio at time n, then

$$\frac{X_n}{(1+r)^n} = \widetilde{\mathbb{E}}_n\left[\frac{X_N}{(1+r)^N}\right], \quad 0 \le n \le N.$$

If we want to have X_N agree with the value V_N of a derivative security at its expiration time N, then we must have

$$\frac{X_n}{(1+r)^n} = \widetilde{\mathbb{E}}_n\left[\frac{X_N}{(1+r)^N}\right] = \widetilde{\mathbb{E}}_n\left[\frac{V_N}{(1+r)^N}\right] \tag{2.4.9}$$

at all times $n = 0, 1, \dots, N$. When a portfolio does this, we define the value V_n of the derivative security at time n to be X_n, and we thus have the risk-neutral pricing formula

$$V_n = \widetilde{\mathbb{E}}_n \left[\frac{V_N}{(1+r)^{N-n}} \right]. \tag{2.4.11}$$

A *Markov process* is a sequence of random variables $X_0, X_1, \ldots, X_N$ with the following property. Suppose n is a time between 0 and $N-1$, we have observed the first n coin tosses $\omega_1 \ldots \omega_n$, and we want to estimate either a function of X_{n+1} or, more generally, a function of X_{n+k} for some k between 1 and $N-n$. We know both the individual coin tosses $\omega_1 \ldots \omega_n$ and the resulting value $X_n(\omega_1 \ldots \omega_n)$ and can base our estimate on this information. For a Markov process, knowledge of the individual coin tosses (the "path") does not provide any information relevant to this estimation problem beyond that information already contained in our knowledge of the value $X_n(\omega_1 \ldots \omega_n)$.

Consider an underlying asset-price process $X_0, X_1, \ldots, X_N$ that is Markov under the risk-neutral measure and a derivative security payoff at time N that is a function of this asset price at time N; i.e., $V_N = v_N(X_N)$. The price of the derivative security at all times n prior to expiration is a function of the underlying asset price at those times; i.e.,

$$V_n = v_n(X_n), \; n = 0, 1, \ldots, N. \tag{2.4.11}$$

In this notation, V_n is a random variable depending on the coin tosses $\omega_1 \ldots \omega_n$. It is potentially path-dependent. On the other hand, $v_n(x)$ is a function of a real number x. When we replace x by the random variable X_n, then $v_n(X_n)$ also becomes random, but in a way that is guaranteed not to be path-dependent. Equation (2.4.11) thus guarantees that the price of the derivative security is not path-dependent.

2.7 Notes

The sample space view of probability theory dates back to Kolmogorov [29], who developed it in a way that extends to infinite probability spaces. We take up this subject in Chapters 1 and 2 of Volume II. Martingales were invented by Doob [13], who attributes the idea and the name "martingale" to a gambling strategy discussed by Ville [43].

The risk-neutral pricing formula is due to Harrison and Kreps [17] and Harrison and Pliska [18].

2.8 Exercises

Exercise 2.1. Using Definition 2.1.1, show the following.

(i) If A is an event and A^c denotes its complement, then $\mathbb{P}(A^c) = 1 - \mathbb{P}(A)$.

(ii) If $A_1, A_2, \ldots, A_N$ is a finite set of events, then

$$\mathbb{P}\left(\bigcup_{n=1}^{N} A_n\right) \leq \sum_{n=1}^{N} \mathbb{P}(A_n). \tag{2.8.1}$$

If the events $A_1, A_2, \ldots, A_N$ are disjoint, then equality holds in (2.8.1).

Exercise 2.2. Consider the stock price S_3 in Figure 2.3.1.

(i) What is the distribution of S_3 under the risk-neutral probabilities $\tilde{p} = \frac{1}{2}$, $\tilde{q} = \frac{1}{2}$.

(ii) Compute $\widetilde{\mathbb{E}}S_1$, $\widetilde{\mathbb{E}}S_2$, and $\widetilde{\mathbb{E}}S_3$. What is the average rate of growth of the stock price under $\widetilde{\mathbb{P}}$?

(iii) Answer (i) and (ii) again under the actual probabilities $p = \frac{2}{3}$, $q = \frac{1}{3}$.

Exercise 2.3. Show that a convex function of a martingale is a submartingale. In other words, let $M_0, M_1, \ldots, M_N$ be a martingale and let φ be a convex function. Show that $\varphi(M_0), \varphi(M_1), \ldots, \varphi(M_N)$ is a submartingale.

Exercise 2.4. Toss a coin repeatedly. Assume the probability of head on each toss is $\frac{1}{2}$, as is the probability of tail. Let $X_j = 1$ if the jth toss results in a head and $X_j = -1$ if the jth toss results in a tail. Consider the stochastic process $M_0, M_1, M_2, \ldots$ defined by $M_0 = 0$ and

$$M_n = \sum_{j=1}^{n} X_j, \; n \geq 1.$$

This is called a *symmetric random walk*; with each head, it steps up one, and with each tail, it steps down one.

(i) Using the properties of Theorem 2.3.2, show that $M_0, M_1, M_2, \ldots$ is a martingale.

(ii) Let σ be a positive constant and, for $n \geq 0$, define

$$S_n = e^{\sigma M_n}\left(\frac{2}{e^{\sigma} + e^{-\sigma}}\right)^n.$$

Show that $S_0, S_1, S_2, \ldots$ is a martingale. Note that even though the symmetric random walk M_n has no tendency to grow, the "geometric symmetric random walk" $e^{\sigma M_n}$ does have a tendency to grow. This is the result of putting a martingale into the (convex) exponential function (see Exercise 2.3). In order to again have a martingale, we must "discount" the geometric symmetric random walk, using the term $\frac{2}{e^{\sigma}+e^{-\sigma}}$ as the discount rate. This term is strictly less than one unless $\sigma = 0$.

Exercise 2.5. Let $M_0, M_1, M_2, \ldots$ be the symmetric random walk of Exercise 2.4, and define $I_0 = 0$ and

$$I_n = \sum_{j=0}^{n-1} M_j(M_{j+1} - M_j), \; n = 1, 2, \ldots.$$

(i) Show that

$$I_n = \frac{1}{2}M_n^2 - \frac{n}{2}.$$

(ii) Let n be an arbitrary nonnegative integer, and let $f(i)$ be an arbitrary function of a variable i. In terms of n and f, define another function $g(i)$ satisfying

$$E_n\big[f(I_{n+1})\big] = g(I_n).$$

Note that although the function $g(I_n)$ on the right-hand side of this equation may depend on n, the only random variable that may appear in its argument is I_n; the random variable M_n may not appear. You will need to use the formula in part (i). The conclusion of part (ii) is that the process $I_0, I_1, I_2, \ldots$ is a Markov process.

Exercise 2.6 (Discrete-time stochastic integral). Suppose $M_0, M_1, \ldots, M_N$ is a martingale, and let $\Delta_0, \Delta_1, \ldots, \Delta_{N-1}$ be an adapted process. Define the *discrete-time stochastic integral* (sometimes called a *martingale transform*) $I_0, I_1, \ldots, I_N$ by setting $I_0 = 0$ and

$$I_n = \sum_{j=0}^{n-1} \Delta_j(M_{j+1} - M_j),\ n = 1, \ldots, N.$$

Show that $I_0, I_1, \ldots, I_N$ is a martingale.

Exercise 2.7. In a binomial model, give an example of a stochastic process that is a martingale but is not Markov.

Exercise 2.8. Consider an N-period binomial model.

(i) Let $M_0, M_1, \ldots, M_N$ and $M_0', M_1', \ldots, M_N'$ be martingales under the risk-neutral measure $\widetilde{\mathbb{P}}$. Show that if $M_N = M_N'$ (for every possible outcome of the sequence of coin tosses), then, for each n between 0 and N, we have $M_n = M_n'$ (for every possible outcome of the sequence of coin tosses).

(ii) Let V_N be the payoff at time N of some derivative security. This is a random variable that can depend on all N coin tosses. Define recursively $V_{N-1}, V_{N-2}, \ldots, V_0$ by the algorithm (1.2.16) of Chapter 1. Show that

$$V_0, \frac{V_1}{1+r}, \ldots, \frac{V_{N-1}}{(1+r)^{N-1}}, \frac{V_N}{(1+r)^N}$$

is a martingale under $\widetilde{\mathbb{P}}$.

(iii) Using the risk-neutral pricing formula (2.4.11) of this chapter, define

$$V_n' = \widetilde{\mathbb{E}}_n\left[\frac{V_N}{(1+r)^{N-n}}\right],\ n = 0, 1, \ldots, N-1.$$

Show that

$$V_0', \frac{V_1'}{1+r}, \ldots, \frac{V_{N-1}'}{(1+r)^{N-1}}, \frac{V_N}{(1+r)^N}$$

is a martingale.

(iv) Conclude that $V_n = V_n'$ for every n (i.e., the algorithm (1.2.16) of Theorem 1.2.2 of Chapter 1 gives the same derivative security prices as the risk-neutral pricing formula (2.4.11) of Chapter 2).

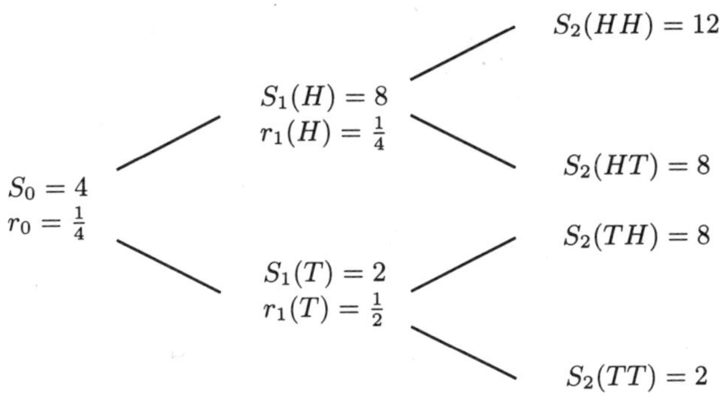

Fig. 2.8.1. A stochastic volatility, random interest rate model.

Exercise 2.9 (Stochastic volatility, random interest rate). Consider a two-period stochastic volatility, random interest rate model of the type described in Exercise 1.9 of Chapter 1. The stock prices and interest rates are shown in Figure 2.8.1.

(i) Determine risk-neutral probabilities

$$\widetilde{\mathbb{P}}(HH),\ \widetilde{\mathbb{P}}(HT),\ \widetilde{\mathbb{P}}(TH),\ \widetilde{\mathbb{P}}(TT),$$

such that the time-zero value of an option that pays off V_2 at time two is given by the risk-neutral pricing formula

$$V_0 = \widetilde{\mathbb{E}}\left[\frac{V_2}{(1+r_0)(1+r_1)}\right].$$

(ii) Let $V_2 = (S_2 - 7)^+$. Compute V_0, $V_1(H)$, and $V_1(T)$.

(iii) Suppose an agent sells the option in (ii) for V_0 at time zero. Compute the position Δ_0 she should take in the stock at time zero so that at time one, regardless of whether the first coin toss results in head or tail, the value of her portfolio is V_1.

(iv) Suppose in (iii) that the first coin toss results in head. What position $\Delta_1(H)$ should the agent now take in the stock to be sure that, regardless of whether the second coin toss results in head or tail, the value of her portfolio at time two will be $(S_2 - 7)^+$?

Exercise 2.10 (Dividend-paying stock). We consider a binomial asset pricing model as in Chapter 1, except that, after each movement in the stock price, a dividend is paid and the stock price is reduced accordingly. To describe this in equations, we define

$$Y_{n+1}(\omega_1 \dots \omega_n \omega_{n+1}) = \begin{cases} u, \text{ if } \omega_{n+1} = H, \\ d, \text{ if } \omega_{n+1} = T. \end{cases}$$

Note that Y_{n+1} depends only on the $(n+1)$st coin toss. In the binomial model of Chapter 1, $Y_{n+1}S_n$ was the stock price at time $n+1$. In the dividend-paying model considered here, we have a random variable $A_{n+1}(\omega_1 \dots \omega_n \omega_{n+1})$, taking values in $(0, 1)$, and the dividend paid at time $n+1$ is $A_{n+1}Y_{n+1}S_n$. After the dividend is paid, the stock price at time $n+1$ is

$$S_{n+1} = (1 - A_{n+1})Y_{n+1}S_n.$$

An agent who begins with initial capital X_0 and at each time n takes a position of Δ_n shares of stock, where Δ_n depends only on the first n coin tosses, has a portfolio value governed by the wealth equation (see (2.4.6))

$$\begin{aligned} X_{n+1} &= \Delta_n S_{n+1} + (1+r)(X_n - \Delta_n S_n) + \Delta_n A_{n+1} Y_{n+1} S_n \\ &= \Delta_n Y_{n+1} S_n + (1+r)(X_n - \Delta_n S_n). \end{aligned} \tag{2.8.2}$$

(i) Show that the discounted wealth process is a martingale under the risk-neutral measure (i.e., Theorem 2.4.5 still holds for the wealth process (2.8.2)). As usual, the risk-neutral measure is still defined by the equations

$$\tilde{p} = \frac{1+r-d}{u-d}, \quad \tilde{q} = \frac{u-1-r}{u-d}.$$

(ii) Show that the risk-neutral pricing formula still applies (i.e., Theorem 2.4.7 holds for the dividend-paying model).

(iii) Show that the discounted stock price is not a martingale under the risk-neutral measure (i.e., Theorem 2.4.4 no longer holds). However, if A_{n+1} is a constant $a \in (0, 1)$, regardless of the value of n and the outcome of the coin tossing $\omega_1 \dots \omega_{n+1}$, then $\frac{S_n}{(1-a)^n(1+r)^n}$ is a martingale under the risk-neutral measure.

Exercise 2.11 (Put–call parity). Consider a stock that pays no dividend in an N-period binomial model. A European call has payoff $C_N = (S_N - K)^+$ at time N. The price C_n of this call at earlier times is given by the risk-neutral pricing formula (2.4.11):

$$C_n = \widetilde{\mathbb{E}}_n\left[\frac{C_N}{(1+r)^{N-n}}\right], \quad n = 0, 1, \dots, N-1.$$

Consider also a put with payoff $P_N = (K - S_N)^+$ at time N, whose price at earlier times is

$$P_n = \widetilde{\mathbb{E}}_n \left[\frac{P_N}{(1+r)^{N-n}} \right], \quad n = 0, 1, \ldots, N-1.$$

Finally, consider a *forward contract* to buy one share of stock at time N for K dollars. The price of this contract at time N is $F_N = S_N - K$, and its price at earlier times is

$$F_n = \widetilde{\mathbb{E}}_n \left[\frac{F_N}{(1+r)^{N-n}} \right], \quad n = 0, 1, \ldots, N-1.$$

(Note that, unlike the call, the forward contract requires that the stock be purchased at time N for K dollars and has a negative payoff if $S_N < K$.)

(i) If at time zero you buy a forward contract and a put, and hold them until expiration, explain why the payoff you receive is the same as the payoff of a call; i.e., explain why $C_N = F_N + P_N$.
(ii) Using the risk-neutral pricing formulas given above for C_n, P_n, and F_n and the linearity of conditional expectations, show that $C_n = F_n + P_n$ for every n.
(iii) Using the fact that the discounted stock price is a martingale under the risk-neutral measure, show that $F_0 = S_0 - \frac{K}{(1+r)^N}$.
(iv) Suppose you begin at time zero with F_0, buy one share of stock, borrowing money as necessary to do that, and make no further trades. Show that at time N you have a portfolio valued at F_N. (This is called a *static replication* of the forward contract. If you sell the forward contract for F_0 at time zero, you can use this static replication to hedge your short position in the forward contract.)
(v) The *forward price* of the stock at time zero is defined to be that value of K that causes the forward contract to have price zero at time zero. The forward price in this model is $(1+r)^N S_0$. Show that, at time zero, the price of a call struck at the forward price is the same as the price of a put struck at the forward price. This fact is called *put–call parity.*
(vi) If we choose $K = (1+r)^N S_0$, we just saw in (v) that $C_0 = P_0$. Do we have $C_n = P_n$ for every n?

Exercise 2.12 (Chooser option). Let $1 \le m \le N-1$ and $K > 0$ be given. A *chooser option* is a contract sold at time zero that confers on its owner the right to receive either a call or a put at time m. The owner of the chooser may wait until time m before choosing. The call or put chosen expires at time N with strike price K. Show that the time-zero price of a chooser option is the sum of the time-zero price of a put, expiring at time N and having strike price K, and a call, expiring at time m and having strike price $\frac{K}{(1+r)^{N-m}}$. (Hint: Use put–call parity (Exercise 2.11).)

Exercise 2.13 (Asian option). Consider an N-period binomial model. An *Asian option* has a payoff based on the average stock price, i.e.,

$$V_N = f\left(\frac{1}{N+1}\sum_{n=0}^{N} S_n\right),$$

where the function f is determined by the contractual details of the option.

(i) Define $Y_n = \sum_{k=0}^{n} S_k$ and use the Independence Lemma 2.5.3 to show that the two-dimensional process (S_n, Y_n), $n = 0, 1, \ldots, N$ is Markov.

(ii) According to Theorem 2.5.8, the price V_n of the Asian option at time n is some function v_n of S_n and Y_n; i.e.,

$$V_n = v_n(S_n, Y_n),\ n = 0, 1, \ldots, N.$$

Give a formula for $v_N(s, y)$, and provide an algorithm for computing $v_n(s, y)$ in terms of v_{n+1}.

Exercise 2.14 (Asian option continued). Consider an N-period binomial model, and let M be a fixed number between 0 and $N-1$. Consider an Asian option whose payoff at time N is

$$V_N = f\left(\frac{1}{N-M}\sum_{n=M+1}^{N} S_n\right),$$

where again the function f is determined by the contractual details of the option.

(i) Define

$$Y_n = \begin{cases} 0, & \text{if } 0 \le n \le M, \\ \sum_{k=M+1}^{n} S_k, & \text{if } M+1 \le n \le N. \end{cases}$$

Show that the two-dimensional process (S_n, Y_n), $n = 0, 1, \ldots, N$ is Markov (under the risk-neutral measure $\widetilde{\mathbb{P}}$).

(ii) According to Theorem 2.5.8, the price V_n of the Asian option at time n is some function v_n of S_n and Y_n, i.e.,

$$V_n = v_n(S_n, Y_n),\ n = 0, 1, \ldots, N.$$

Of course, when $n \le M$, Y_n is not random and does not need to be included in this function. Thus, for such n we should seek a function v_n of S_n alone and have

$$V_n = \begin{cases} v_n(S_n), & \text{if } 0 \le n \le M, \\ v_n(S_n, Y_n), & \text{if } M+1 \le n \le N. \end{cases}$$

Give a formula for $v_N(s, y)$, and provide an algorithm for computing v_n in terms of v_{n+1}. Note that the algorithm is different for $n < M$ and $n > M$, and there is a separate transition formula for $v_M(s)$ in terms of $v_{M+1}(\cdot, \cdot)$.

3

State Prices

3.1 Change of Measure

In the binomial no-arbitrage pricing model of Chapter 1 and also in the continuous-time models formulated in Chapters 4 and 5 of Volume II, there are two probability measures that merit our attention. One is the *actual probability measure*, by which we mean the one that we seek by empirical estimation of the model parameters. The other is the *risk-neutral probability measure*, under which the discounted prices of assets are martingales. These two probability measures give different weights to the asset-price paths in the model. They agree, however, on which price paths are possible (i.e., which paths have positive probability of occurring); they disagree only on what these positive probabilities are. The actual probabilities are the "right" ones. The risk-neutral probabilities are a fictitious but helpful construct because they allow us to neatly summarize the result of solving systems of equations (see, e.g., the system (1.1.3), (1.1.4) of Chapter 1, which leads to the formula (1.1.7) of that chapter).

Let us more generally consider a finite sample space Ω on which we have two probability measures $\mathbb{P}$ and $\widetilde{\mathbb{P}}$. Let us assume that $\mathbb{P}$ and $\widetilde{\mathbb{P}}$ both give positive probability to every element of Ω, so we can form the quotient

$$Z(\omega) = \frac{\widetilde{\mathbb{P}}(\omega)}{\mathbb{P}(\omega)}. \tag{3.1.1}$$

Because it depends on the outcome ω of a random experiment, Z is a random variable. It is called the *Radon-Nikodým derivative of* $\widetilde{\mathbb{P}}$ *with respect to* $\mathbb{P}$, although in this context of a finite sample space Ω, it is really a quotient rather than a derivative. The random variable Z has three important properties, which we state as a theorem.

Theorem 3.1.1. *Let* $\mathbb{P}$ *and* $\widetilde{\mathbb{P}}$ *be probability measures on a finite sample space* Ω, *assume that* $\mathbb{P}(\omega) > 0$ *and* $\widetilde{\mathbb{P}}(\omega) > 0$ *for every* $\omega \in \Omega$, *and define the random variable* Z *by (3.1.1). Then we have the following:*

(i) $\mathbb{P}(Z > 0) = 1$;
(ii) $\mathbb{E}Z = 1$;
(iii) for any random variable Y,

$$\widetilde{\mathbb{E}}Y = \mathbb{E}[ZY]. \tag{3.1.2}$$

PROOF: Property (i) follows immediately from the fact that we have assumed $\widetilde{\mathbb{P}}(\omega) > 0$ for every $\omega \in \Omega$. Property (ii) can be verified by the computation

$$\mathbb{E}Z = \sum_{\omega \in \Omega} Z(\omega)\mathbb{P}(\omega) = \sum_{\omega \in \Omega} \frac{\widetilde{\mathbb{P}}(\omega)}{\mathbb{P}(\omega)}\mathbb{P}(\omega) = \sum_{\omega \in \Omega} \widetilde{\mathbb{P}}(\omega) = 1,$$

the last equality following from the fact that $\widetilde{\mathbb{P}}$ is a probability measure. The following similar computation verifies property (iii):

$$\begin{aligned}\widetilde{\mathbb{E}}Y &= \sum_{\omega \in \Omega} Y(\omega)\widetilde{\mathbb{P}}(\omega) = \sum_{\omega \in \Omega} Y(\omega)\frac{\widetilde{\mathbb{P}}(\omega)}{\mathbb{P}(\omega)}\mathbb{P}(\omega) \\ &= \sum_{\omega \in \Omega} Y(\omega)Z(\omega)\mathbb{P}(\omega) = \mathbb{E}[ZY]. \qquad \square\end{aligned}$$

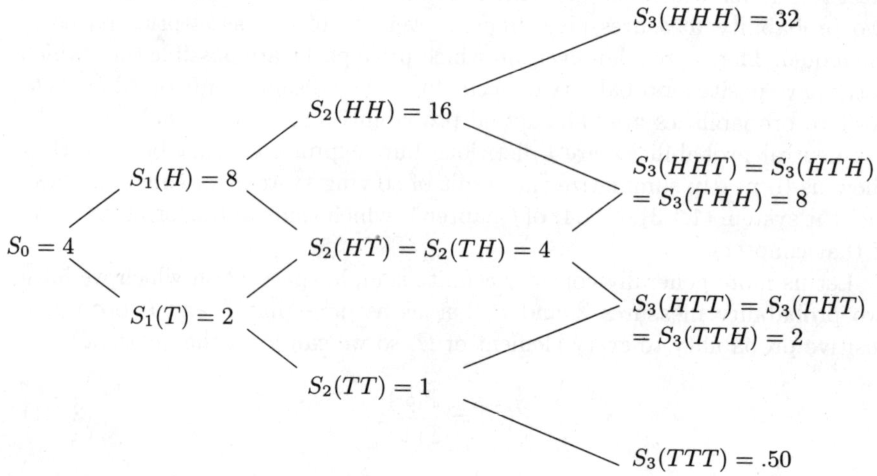

Fig. 3.1.1. A three-period model.

Example 3.1.2. Consider again the three-period model of Figure 3.1.1. The underlying probability space is

$$\Omega = \{HHH, HHT, HTH, HTT, THH, THT, TTH, TTT\}.$$

We take $p = \frac{2}{3}$ as the actual probability of a head and $q = \frac{1}{3}$ as the actual probability of a tail. Then the actual probability measure is

$$\begin{aligned}&\mathbb{P}(HHH) = \frac{8}{27}, \quad \mathbb{P}(HHT) = \frac{4}{27}, \quad \mathbb{P}(HTH) = \frac{4}{27}, \quad \mathbb{P}(HTT) = \frac{2}{27},\\ &\mathbb{P}(THH) = \frac{4}{27}, \quad \mathbb{P}(THT) = \frac{2}{27}, \quad \mathbb{P}(TTH) = \frac{2}{27}, \quad \mathbb{P}(TTT) = \frac{1}{27}.\end{aligned} \tag{3.1.3}$$

We take the interest rate to be $r = \frac{1}{4}$, and then the risk-neutral probability of a head is $\tilde{p} = \frac{1}{2}$ and the risk-neutral probability of a tail is $\tilde{q} = \frac{1}{2}$. The risk-neutral probability measure is

$$\begin{aligned}&\widetilde{\mathbb{P}}(HHH) = \frac{1}{8}, \quad \widetilde{\mathbb{P}}(HHT) = \frac{1}{8}, \quad \widetilde{\mathbb{P}}(HTH) = \frac{1}{8}, \quad \widetilde{\mathbb{P}}(HTT) = \frac{1}{8},\\ &\widetilde{\mathbb{P}}(THH) = \frac{1}{8}, \quad \widetilde{\mathbb{P}}(THT) = \frac{1}{8}, \quad \widetilde{\mathbb{P}}(TTH) = \frac{1}{8}, \quad \widetilde{\mathbb{P}}(TTT) = \frac{1}{8}.\end{aligned} \tag{3.1.4}$$

Therefore, the Radon-Nikodým derivative of $\widetilde{\mathbb{P}}$ with respect to $\mathbb{P}$ is

$$\begin{aligned}&Z(HHH) = \frac{27}{64}, \quad Z(HHT) = \frac{27}{32}, \quad Z(HTH) = \frac{27}{32}, \quad Z(HTT) = \frac{27}{16},\\ &Z(THH) = \frac{27}{32}, \quad Z(THT) = \frac{27}{16}, \quad Z(TTH) = \frac{27}{16}, \quad Z(TTT) = \frac{27}{8}.\end{aligned} \tag{3.1.5}$$

In Example 1.2.4 of Chapter 1, for this model we determined the time-zero price of a lookback option whose payoff at time three was given by

$$\begin{aligned}&V_3(HHH) = 0, \quad V_3(HHT) = 8, \quad V_3(HTH) = 0, \quad V_3(HTT) = 6,\\ &V_3(THH) = 0, \quad V_3(THT) = 2, \quad V_3(TTH) = 2, \quad V_3(TTT) = 3.50.\end{aligned}$$

According to the risk-neutral pricing formula (2.4.11) of Chapter 2, this time-zero value is

$$\begin{aligned}V_0 &= \widetilde{\mathbb{E}}\frac{V_3}{(1+r)^3}\\ &= \left(\frac{4}{5}\right)^3 \sum_{\omega\in\Omega} V_3(\omega)\widetilde{\mathbb{P}}(\omega)\\ &= 0.512\left[0\cdot\frac{1}{8} + 8\cdot\frac{1}{8} + 0\cdot\frac{1}{8} + 6\cdot\frac{1}{8} + 0\cdot\frac{1}{8} + 2\cdot\frac{1}{8} + 2\cdot\frac{1}{8} + 3.50\cdot\frac{1}{8}\right]\\ &= 1.376,\end{aligned} \tag{3.1.6}$$

which is the number determined in Example 1.2.4 of Chapter 1 to be the cost at time zero of setting up a replicating portfolio. Using the random variable Z, we can rewrite (3.1.6) as

$$\begin{aligned}
V_0 &= \mathbb{E}\frac{V_3 Z}{(1+r)^3} \\
&= \left(\frac{4}{5}\right)^3 \sum_{\omega\in\Omega} V_3(\omega)Z(\omega)\mathbb{P}(\omega) \\
&= 0.512\Big[0\cdot\frac{27}{64}\cdot\frac{8}{27} + 8\cdot\frac{27}{32}\cdot\frac{4}{27} + 0\cdot\frac{27}{32}\cdot\frac{4}{27} + 6\cdot\frac{27}{16}\cdot\frac{2}{27} \\
&\qquad + 0\cdot\frac{27}{32}\cdot\frac{4}{27} + 2\cdot\frac{27}{16}\cdot\frac{2}{27} + 2\cdot\frac{27}{16}\cdot\frac{2}{27} + 3.50\cdot\frac{27}{8}\cdot\frac{1}{27}\Big] \\
&= 1.376,
\end{aligned} \tag{3.1.7}$$

The advantage of (3.1.7) over (3.1.6) is that (3.1.7) makes no reference to the risk-neutral measure. However, it does not simply compute the expected discounted payoff of the option under the actual probability measure but rather first weights these payoffs using the random variable Z. This leads to the idea of *state prices*, which we formalize in the next definition. □

Definition 3.1.3. *In the N-period binomial model with actual probability measure $\mathbb{P}$ and risk-neutral probability measure $\widetilde{\mathbb{P}}$, let Z denote the Radon-Nikodým derivative of $\widetilde{\mathbb{P}}$ with respect to $\mathbb{P}$; i.e.,*

$$Z(\omega_1\ldots\omega_N) = \frac{\widetilde{\mathbb{P}}(\omega_1\ldots\omega_N)}{\mathbb{P}(\omega_1\ldots\omega_N)} = \left(\frac{\tilde{p}}{p}\right)^{\#H(\omega_1\ldots\omega_N)}\left(\frac{\tilde{q}}{q}\right)^{\#T(\omega_1\ldots\omega_N)}, \tag{3.1.8}$$

where $\#H(\omega_1\ldots\omega_N)$ denotes the number of heads appearing in the sequence $\omega_1\ldots\omega_N$ and $\#T(\omega_1\ldots\omega_N)$ denotes the number of tails appearing in this sequence. The state price density *random variable is*

$$\zeta(\omega) = \frac{Z(\omega)}{(1+r)^N}, \tag{3.1.9}$$

and $\zeta(\omega)\mathbb{P}(\omega)$ is called the state price *corresponding to ω.*

Let $\overline{\omega} = \overline{\omega}_1\ldots\overline{\omega}_N$ be a particular coin toss sequence in the N-period model, and consider a derivative security that pays off 1 if $\overline{\omega}$ occurs and otherwise pays off 0; i.e.,

$$V_N(\omega) = \begin{cases} 1, & \text{if } \omega = \overline{\omega}, \\ 0, & \text{otherwise.} \end{cases}$$

According to the risk-neutral pricing formula, the value of this derivative security at time zero is

$$\widetilde{\mathbb{E}}\frac{V_N}{(1+r)^N} = \frac{\widetilde{\mathbb{P}}(\overline{\omega})}{(1+r)^N} = \frac{Z(\overline{\omega})\mathbb{P}(\overline{\omega})}{(1+r)^N} = \zeta(\overline{\omega})\mathbb{P}(\overline{\omega}).$$

We see that the state price $\zeta(\overline{\omega})\mathbb{P}(\overline{\omega})$ tells the price at time zero of a contract that pays 1 at time N if and only if $\overline{\omega}$ occurs. This price should include a discount from time N to time zero to account for the time value of money, and the term $\frac{1}{(1+r)^N}$ does indeed appear in (3.1.9). It is natural to expect the price to take into account the probability that $\overline{\omega}$ will occur, and therefore we have arranged the formulas so that $\mathbb{P}(\overline{\omega})$ is one of the factors in the state price. However, these two factors alone cannot tell the whole story because they do not account for risk. If we were to use these terms alone, and take the time-zero price of a derivative security to be $\mathbb{E}\frac{V_N}{(1+r)^N}$, then the time-zero price of an asset would depend only on its expected return under the actual probability measure. In fact, the price of an asset depends on both its expected return and the risk it presents. The remaining term appearing in the state price corresponding to $\overline{\omega}$, $Z(\overline{\omega})$, accounts for risk. For example, in (3.1.5) we see that Z discounts the importance of the stock price paths that end above the initial stock price $S_0 = 4$ because $Z < 1$ whenever there are two or three heads in the three coin tosses, but Z inflates the importance of the stock price paths that end below the initial stock price. The effect of this is to make holding the stock appear less favorable than one would infer from simply computing $\mathbb{E}\left[\left(\frac{4}{5}\right)^3 S_3\right]$, its discounted expected value at time three.

The state price $\zeta(\overline{\omega})\mathbb{P}(\overline{\omega})$ tells us the time-zero price of a contract that pays 1 at time N if and only if $\overline{\omega}$ occurs. The state price density $\frac{Z(\overline{\omega})}{(1+r)^N}$ tells us the time-zero price of this contract per unit of actual probability. For this reason, we call it a density.

Of course, most contracts make payoffs for several different values of ω, and these payoffs are not all necessarily 1. Such a contract can be regarded as a portfolio of simple contracts, each of which pays off 1 if and only if some particular ω occurs, and their prices can be computed by summing up the prices of these components. To see this, recall from (2.4.11) of Chapter 2 the risk-neutral pricing formula $V_0 = \widetilde{\mathbb{E}}\frac{V_N}{(1+r)^N}$ for the time-zero price of an arbitrary derivative security paying V_N at time N. In terms of the state price density, this can be rewritten simply as

$$V_0 = \mathbb{E}[\zeta V_N] = \sum_{\omega\in\Omega} V_N(\omega)\zeta(\omega)\mathbb{P}(\omega). \tag{3.1.10}$$

Equation (3.1.7) is a special case of this, where the $\zeta(\omega)$ term is separated into its factors $\frac{1}{(1+r)^3}$ and $Z(\omega)$.

3.2 Radon-Nikodým Derivative Process

In the previous section, we considered the Radon-Nikodým derivative of the risk-neutral probability measure with respect to the actual probability measure in an N-period binomial model. This random variable Z depends on the N coin tosses in the model. To get related random variables that depend on fewer coin tosses, we can estimate Z based on the information at time $n < N$.

This procedure of estimation will occur in other contexts as well, and thus we give a general result that does not require that Z is a Radon-Nikodým derivative.

Theorem 3.2.1. *Let Z be a random variable in an N-period binomial model. Define*

$$Z_n = \mathbb{E}_n Z,\ n = 0, 1, \ldots, N. \tag{3.2.1}$$

Then Z_n, $n = 0, 1, \ldots, N$, is a martingale under $\mathbb{P}$.

PROOF: For $n = 0, 1, \ldots, N-1$, we use the "iterated conditioning" property of Theorem 2.3.2(iii) of Chapter 2 to compute

$$\mathbb{E}_n[Z_{n+1}] = \mathbb{E}_n[\mathbb{E}_{n+1}[Z]] = \mathbb{E}_n[Z] = Z_n.$$

This shows that Z_n, $n = 0, 1, \ldots, N$ is a martingale. □

Remark 3.2.2. Although Theorem 3.2.1 is stated for the probability measure $\mathbb{P}$, the analogous theorem is true under the risk-neutral probability measure $\widetilde{\mathbb{P}}$. The proof is the same. □

When successive estimates of a random variable are made, the estimates become more precise with increasing time (and information). However, Theorem 3.2.1 says they have no tendency to rise or fall. If a later estimate were on average higher than an earlier estimate, this tendency to rise would have already been incorporated into the earlier estimate. This is similar to the situation with an efficient stock market. If a stock were known to outperform other stocks having the same level of risk, this fact would have already been incorporated into the current price of the stock and thereby raise it to the point where the superior performance was no longer possible.

Example 3.2.3. Consider the three-period model of Example 3.1.2. In that example, we determined the Radon-Nikodým derivative of $\widetilde{\mathbb{P}}$ with respect to $\mathbb{P}$ to be given by (3.1.5). For $n = 0, 1, 2, 3$, we define $Z_n = \mathbb{E}_n[Z]$. In particular, $Z_3(\omega_1\omega_2\omega_3) = Z(\omega_1\omega_2\omega_3)$ for all $\omega_1\omega_2\omega_3$. We compute

$$Z_2(HH) = \frac{2}{3} Z_3(HHH) + \frac{1}{3} Z_3(HHT) = \frac{9}{16},$$

$$Z_2(HT) = \frac{2}{3} Z_3(HTH) + \frac{1}{3} Z_2(HTT) = \frac{9}{8},$$

$$Z_2(TH) = \frac{2}{3} Z_3(THH) + \frac{1}{3} Z_2(THT) = \frac{9}{8},$$

$$Z_2(TT) = \frac{2}{3} Z_2(TTH) + \frac{1}{3} Z^2(TTT) = \frac{9}{4}.$$

According to its definition, $Z_1 = \mathbb{E}_1[Z]$, but Theorem 3.2.1 allows us to compute it using the martingale formula $Z_1 = \mathbb{E}_1[Z_2]$. This leads to the equations

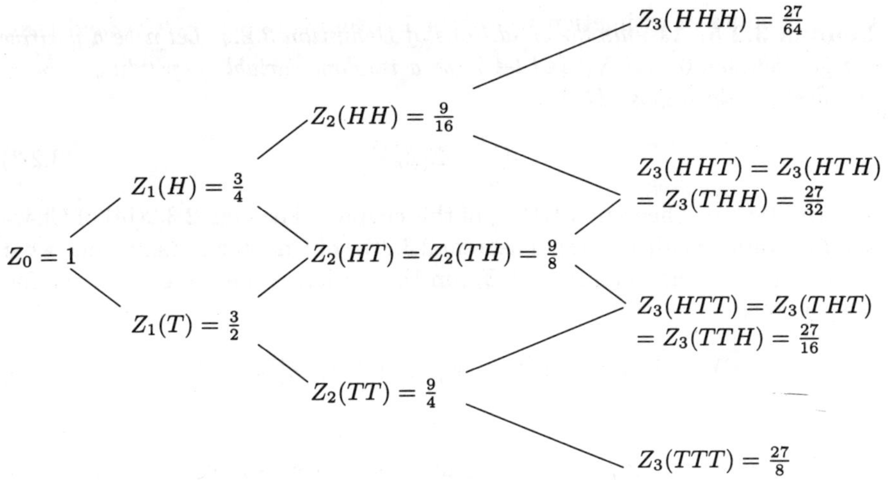

Fig. 3.2.1. A Radon-Nikodým derivative process.

$$Z_1(H) = \frac{2}{3} Z_2(HH) + \frac{1}{3} Z_2(HT) = \frac{3}{4},$$

$$Z_1(T) = \frac{2}{3} Z_1(TH) + \frac{1}{3} Z_1(TT) = \frac{3}{2}.$$

According to its definition, $Z_0 = \mathbb{E}Z$, which must be 1 because of Theorem 3.1.1(ii). We can also compute it using the martingale formula $Z_0 = \mathbb{E}_0[Z_1] = \mathbb{E}Z_1$, and this leads to

$$Z_0 = \tfrac{2}{3} Z_1(H) + \tfrac{1}{3} Z_1(T) = 1.$$

The process Z_n, $n = 0, 1, 2, 3$, is shown in Figure 3.2.1. □

Definition 3.2.4. *In an N-period binomial model, let $\mathbb{P}$ be the actual probability measure, $\widetilde{\mathbb{P}}$ the risk-neutral probability measure, and assume that $\mathbb{P}(\omega) > 0$ and $\widetilde{\mathbb{P}}(\omega) > 0$ for every sequence of coin tosses ω. Define the Radon-Nikodým derivative (random variable) $Z(\omega) = \frac{\widetilde{\mathbb{P}}(\omega)}{\mathbb{P}(\omega)}$ for every ω. The* Radon-Nikodým derivative process *is*

$$Z_n = \mathbb{E}_n[Z], \; n = 0, 1, \ldots, N. \tag{3.2.2}$$

In particular, $Z_N = Z$ and $Z_0 = 1$.

In the context of Definition 3.2.4, we can compute the risk-neutral expectation of a random variable Y by computing under the actual probability measure the expectation $E[ZY]$. If Y only depends on the first n coin tosses, where $n < N$, this computation can be simplified further.

Lemma 3.2.5. *Assume the conditions of Definition 3.2.4. Let n be a positive integer between 0 and N, and let Y be a random variable depending only on the first n coin tosses. Then*

$$\widetilde{\mathbb{E}}Y = \mathbb{E}[Z_n Y]. \tag{3.2.3}$$

PROOF: We use Theorem 3.1.1(iii) of this chapter, Theorem 2.3.2(iii) of Chapter 2 (iterated conditioning), Theorem 2.3.2(ii) of Chapter 2 (taking out what is known), and the definition of Z_n, in that order, to justify each of the following steps:

$$\widetilde{\mathbb{E}}Y = \mathbb{E}\big[ZY\big] = \mathbb{E}\big[\mathbb{E}_n[ZY]\big] = \mathbb{E}\big[Y\mathbb{E}_n[Z]\big] = \mathbb{E}\big[YZ_n\big]. \qquad \square$$

An illuminating application of Lemma 3.2.5 occurs if we fix a sequence of n coin tosses, $\overline{\omega}_1 \dots \overline{\omega}_n$, and define

$$Y(\omega_1 \dots \omega_n \omega_{n+1} \dots \omega_N) = \begin{cases} 1, \text{ if } \omega_1 \dots \omega_n = \overline{\omega}_1 \dots \overline{\omega}_n, \\ 0, \text{ otherwise.} \end{cases}$$

In other words, Y takes the value 1 if and only if the first n coin tosses result in the particular sequence $\overline{\omega}_1 \dots \overline{\omega}_n$ we have fixed in advance. The coin tosses $\omega_{n+1} \dots \omega_N$ are irrelevant. Then

$$\begin{aligned} \widetilde{\mathbb{E}}Y &= \widetilde{\mathbb{P}}\{\text{The first } n \text{ coin tosses result in } \overline{\omega}_1 \dots \overline{\omega}_n\} \\ &= \tilde{p}^{\#H(\overline{\omega}_1 \dots \overline{\omega}_n)} \tilde{q}^{\#T(\overline{\omega}_1 \dots \overline{\omega}_n)}, \end{aligned}$$

where the notation $\#H(\cdots)$ and $\#T(\cdots)$ is explained in Definition 3.1.3. On the other hand,

$$\begin{aligned} \mathbb{E}[YZ_n] &= Z_n(\overline{\omega}_1 \dots \overline{\omega}_n)\mathbb{P}\{\text{The first } n \text{ coin tosses result in } \overline{\omega}_1 \dots \overline{\omega}_n\} \\ &= Z_n(\overline{\omega}_1 \dots \overline{\omega}_n) p^{\#H(\overline{\omega}_1 \dots \overline{\omega}_n)} q^{\#T(\overline{\omega}_1 \dots \overline{\omega}_n)}. \end{aligned}$$

Lemma 3.2.5 asserts that these two quantities are equal, and hence

$$Z_n(\overline{\omega}_1 \dots \overline{\omega}_n) = \left(\frac{\tilde{p}}{p}\right)^{\#H(\overline{\omega}_1 \dots \overline{\omega}_n)} \left(\frac{\tilde{q}}{q}\right)^{\#T(\overline{\omega}_1 \dots \overline{\omega}_n)} \tag{3.2.4}$$

This can be verified in Figure 3.2.1. For example, in that figure we have

$$Z_2(HH) = \left(\frac{\tilde{p}}{p}\right)^2 \left(\frac{\tilde{q}}{q}\right)^0 = \left(\frac{\frac{1}{2}}{\frac{2}{3}}\right)^2 = \frac{9}{16}$$

and

$$Z_3(HTT) = \left(\frac{\tilde{p}}{p}\right)^1 \left(\frac{\tilde{q}}{q}\right)^2 = \left(\frac{\frac{1}{2}}{\frac{2}{3}}\right)^1 \left(\frac{\frac{1}{2}}{\frac{1}{3}}\right)^2 = \frac{27}{16}.$$

We see that, for each n, $Z_n(\omega_1 \dots \omega_n)$ is the ratio of the risk-neutral probability and the actual probability of obtaining the sequence of coin tosses $\omega_1 \dots \omega_n$. Lemma 3.2.5 asserts that if Y depends only on the first n coin tosses, then we do not need to consider the coin tosses after time n. We may use Z_n as a surrogate for the Radon-Nikodým derivative Z in the formula $\widetilde{\mathbb{E}}Y = \mathbb{E}[ZY]$ of Theorem 3.1.1(iii), and Z_n is computed just like Z, except that Z_n is a ratio of probabilities for the first n coin tosses rather than all N tosses.

In addition to relating expectations under the two measures $\mathbb{P}$ and $\widetilde{\mathbb{P}}$, we want to have a formula relating conditional expectations under these measures. This is provided by the following lemma.

Lemma 3.2.6. *Assume the conditions of Definition 3.2.4. Let $n \leq m$ be positive integers between 0 and N, and let Y be a random variable depending only on the first m coin tosses. Then*

$$\widetilde{\mathbb{E}}_n[Y] = \frac{1}{Z_n}\mathbb{E}_n[Z_m Y]. \tag{3.2.5}$$

PROOF: Let $\omega_1 \dots \omega_n$ be given. We compute

$$\begin{aligned}
&\widetilde{\mathbb{E}}_n[Y](\omega_1 \dots \omega_n)\\
&= \sum_{\omega_{n+1}\dots\omega_m} Y(\omega_1 \dots \omega_m)\tilde{p}^{\#H(\omega_{n+1}\dots\omega_m)}\tilde{q}^{\#T(\omega_{n+1}\dots\omega_m)}\\
&= \left(\frac{p}{\tilde{p}}\right)^{\#H(\omega_1\dots\omega_n)}\left(\frac{q}{\tilde{q}}\right)^{\#T(\omega_1\dots\omega_n)}\\
&\quad \cdot \sum_{\omega_{n+1}\dots\omega_m}\Bigg[Y(\omega_1 \dots \omega_m)\left(\frac{\tilde{p}}{p}\right)^{\#H(\omega_1\dots\omega_m)}\left(\frac{\tilde{q}}{q}\right)^{\#T(\omega_1\dots\omega_m)}\\
&\qquad\qquad \cdot p^{\#H(\omega_{n+1}\dots\omega_m)}q^{\#T(\omega_{n+1}\dots\omega_m)}\Bigg]\\
&= \frac{1}{Z_n(\omega_1 \dots \omega_n)}\\
&\quad \cdot \sum_{\omega_{n+1}\dots\omega_m} Y(\omega_1 \dots \omega_m)Z_m(\omega_1 \dots \omega_m)p^{\#H(\omega_{n+1}\dots\omega_m)}q^{\#T(\omega_{n+1}\dots\omega_m)}\\
&= \frac{1}{Z_n(\omega_1 \dots \omega_n)}\mathbb{E}_n[YZ_m](\omega_1 \dots \omega_n).
\end{aligned}$$

□

We are now in a position to give a variety of formulations of the risk-neutral pricing formula.

Theorem 3.2.7. *Consider an N-period binomial model with $0 < d < 1+r < u$. Assume that the actual probability for head, p, and the actual probability for tail, q, are positive. The risk-neutral probabilities for head and tail are given, as usual, by*

$$\tilde{p} = \frac{1+r-d}{u-d}, \quad \tilde{q} = \frac{u-1-r}{u-d},$$

and these also are both positive. Let $\mathbb{P}$ and $\widetilde{\mathbb{P}}$ denote the corresponding actual and risk-neutral probability measures, respectively, let Z be the Radon-Nikodým derivative of $\widetilde{\mathbb{P}}$ with respect to $\mathbb{P}$, and let Z_n, $n = 0, 1, \ldots, N$, be the Radon-Nikodým derivative process.

Consider a derivative security whose payoff V_N may depend on all N coin tosses. For $n = 0, 1, \ldots, N$, the price at time n of this derivative security is

$$V_n = \widetilde{\mathbb{E}}_n \frac{V_N}{(1+r)^{N-n}} = \frac{(1+r)^n}{Z_n}\mathbb{E}_n \frac{Z_N V_N}{(1+r)^N} = \frac{1}{\zeta_n}\mathbb{E}_n[\zeta_N V_N], \tag{3.2.6}$$

where the state price density process ζ_n *is defined by*

$$\zeta_n = \frac{Z_n}{(1+r)^n}, \quad n = 0, 1, \ldots, N. \tag{3.2.7}$$

PROOF: The first equality in (3.2.6) is (2.4.11) of Chapter 2. The second equality follows from Lemma 3.2.6. The third is just a matter of definition of ζ_n. □

3.3 Capital Asset Pricing Model

The no-arbitrage pricing methodology of this text is one of two different ways of modeling prices of assets. The other, the *capital asset pricing model*, is based on balancing supply with demand among investors who have utility functions that convert units of consumption to units of satisfaction. The capital asset pricing model provides useful qualitative insights into markets but does not yield the precise quantitative results available through the no-arbitrage methodology. Moreover, in an idealized complete market, the no-arbitrage argument is compelling. On the other hand, many markets are incomplete, and prices cannot be determined from no-arbitrage considerations alone. Utility-based models are still the only theoretically defensible way of treating such markets, although there is a widespread practice of using "risk-neutral" pricing, even when the assets being priced cannot be replicated by trading in other, more primitive assets.

This text is about no-arbitrage pricing in complete markets and the mathematical methodology that supports this point of view. The mathematical methodology, however, is broadly applicable. In this section, we show how it can be brought to bear on a problem at the heart of the capital asset pricing model, that of maximizing the expected utility obtained from investment.

In no-arbitrage pricing, there are two probability measures, the actual probability measure and the risk-neutral measure. When pricing derivative securities, we need only consider the risk-neutral measure. There are, however, two situations in which the actual probability measure becomes relevant: asset management and risk management. In asset management, one cares about the trade-off of risk and actual (rather than risk-neutral) expected return. In risk management, one cares about the actual probability of a catastrophic event. In both of these situations, however, there is a role for the risk-neutral probability measure. For risk management, the portfolio whose risk is being assessed normally contains derivative securities whose theoretical prices under various scenarios must be computed using the risk-neutral measure. For asset management, the risk-neutral measure enters in the manner set forth in this section.

We now set out the capital asset pricing problem. By a *utility function* we shall mean a nondecreasing, concave function defined on the set of real numbers. This function may take the value $-\infty$, but not the value $+\infty$. A common utility function is $\ln x$, which is normally defined only for $x > 0$. We adopt the convention that $\ln x = -\infty$ for $x \leq 0$, so this is defined for every $x \in \mathbb{R}$ and is nondecreasing and concave. Recall that a function U is concave if

$$U(\alpha x + (1-\alpha)y) \geq \alpha U(x) + (1-\alpha)U(y) \quad \text{for every } x, y \in \mathbb{R},\ \alpha \in (0,1). \tag{3.3.1}$$

We say U is *strictly concave* if the inequality in (3.3.1) is strict whenever $x \neq y$, and in fact we shall assume that U is strictly concave everywhere it is finite. A whole class of utility functions can be obtained by first choosing a number $p < 1$, $p \neq 0$, and another number $c \in \mathbb{R}$, and defining

$$U_p(x) = \begin{cases} \frac{1}{p}(x-c)^p, \text{ if } x > c, \\ 0, & \text{if } 0 < p < 1 \text{ and } x = c, \\ -\infty, & \text{if } p < 0 \text{ and } x = c, \\ -\infty, & \text{if } x < c. \end{cases}$$

For these functions, the *index of absolute risk aversion* $-\frac{U''(x)}{U'(x)}$ is the hyperbolic function $\frac{1-p}{x-c}$ for $x > c$. This class of functions is called the *HARA* (*h*yperbolic *a*bsolute *r*isk *a*version) class. The HARA function corresponding to $p = 0$ is

$$U_0(x) = \begin{cases} \ln(x-c), & \text{if } x > c, \\ -\infty, & \text{if } x \leq c. \end{cases}$$

Concavity of utility functions is assumed in order to capture the trade-off between risk and return. For example, consider a gamble which pays off 1 with probability $\frac{1}{2}$ and 99 with probability $\frac{1}{2}$. The expected payoff is 50, but a risk-averse agent would prefer to have 50 rather than the random payoff of the gamble. Let X denote this random payoff, i.e., $\mathbb{P}(X=1) = \mathbb{P}(X=99) = \frac{1}{2}$. For a concave utility function, we have from Jensen's inequality used upside down (Theorem 2.2.5 of Chapter 2) that $\mathbb{E}U(X) \leq U(\mathbb{E}X)$. Indeed, if $U(x) = \ln x$, then $\mathbb{E}\ln X = 2.30$ and $\ln \mathbb{E}X = 3.91$. If we model agent behavior as maximization of expected utility of payoff, our model would indicate that an agent would prefer the nonrandom payoff $\mathbb{E}X = 50$ over the random payoff X.

By comparing expected utility of payoffs rather than expected payoffs, and choosing the utility function judiciously, we can capture an investor's attitude toward the trade-off between risk and return.

Let us consider an N-period binomial model with the usual parameters $0 < d < 1+r < u$. An agent begins with initial wealth X_0 and wishes to invest in the stock and the money market account so as to maximize the expected utility of his wealth at time N. In other words, the agent has a utility function U and wishes to solve the following problem.

Problem 3.3.1 (Optimal investment). Given X_0, find an adapted portfolio process $\Delta_0, \Delta_1, \ldots, \Delta_{N-1}$ that maximizes

$$\mathbb{E}U(X_N) \tag{3.3.2}$$

subject to the wealth equation

$$X_{n+1} = \Delta_n S_{n+1} + (1+r)(X_n - \Delta_n S_n),\ n = 0, 1, \ldots, N-1.$$

Note that the expectation in (3.3.2) is computed using the actual probability measure $\mathbb{P}$. The agent is risk-averse and uses his utility function U to capture the trade-off between actual risk and actual return. It does not make sense to do this under the risk-neutral measure because under the risk-neutral measure both the stock and the money market account have the same expected rate of return; an agent seeking to maximize $\widetilde{\mathbb{E}}U(X)$ would invest only in the money market.

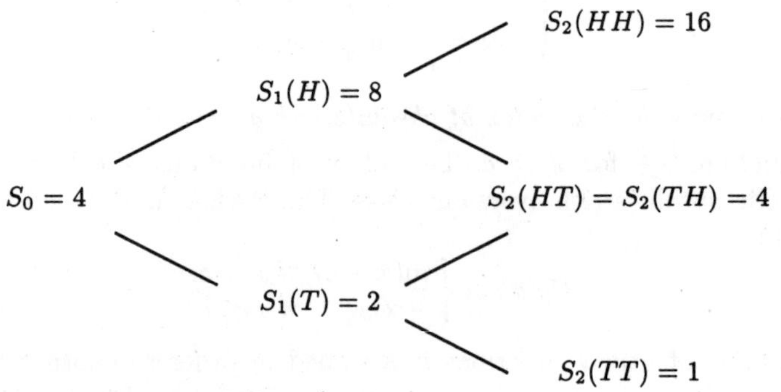

Fig. 3.3.1. A two-period model.

Example 3.3.2. Consider the two-period model of Figure 3.3.1, in which the interest rate is $r = \frac{1}{4}$, so the risk-neutral probability measure is

$$\widetilde{\mathbb{P}}(HH)=\frac{1}{4},\ \widetilde{\mathbb{P}}(HT)=\frac{1}{4},\ \widetilde{\mathbb{P}}(TH)=\frac{1}{4},\ \widetilde{\mathbb{P}}(TT)=\frac{1}{4}.$$

Assume the actual probability measure is

$$\mathbb{P}(HH)=\frac{4}{9},\ \mathbb{P}(HT)=\frac{2}{9},\ \mathbb{P}(TH)=\frac{2}{9},\ \mathbb{P}(TT)=\frac{1}{9}.$$

Consider an agent who begins with $X_0=4$ and wants to choose Δ_0, $\Delta_1(H)$ and $\Delta_1(T)$ in order to maximize $\mathbb{E}\ln X_2$. Note that

$$X_1(H)=8\Delta_0+\frac{5}{4}(4-4\Delta_0)=3\Delta_0+5,$$

$$X_1(T)=2\Delta_0+\frac{5}{4}(4-4\Delta_0)=-3\Delta_0+5,$$

and

$$\begin{aligned}X_2(HH)&=16\Delta_1(H)+\frac{5}{4}(X_1(H)-8\Delta_1(H))\\&=6\Delta_1(H)+\frac{15}{4}\Delta_0+\frac{25}{4},\end{aligned}\tag{3.3.3}$$

$$\begin{aligned}X_2(HT)&=4\Delta_1(H)+\frac{5}{4}(X_1(H)-8\Delta_1(H))\\&=-6\Delta_1(H)+\frac{15}{4}\Delta_0+\frac{25}{4},\end{aligned}\tag{3.3.4}$$

$$\begin{aligned}X_2(TH)&=4\Delta_1(T)+\frac{5}{4}(X_1(T)-2\Delta_1(T))\\&=\frac{3}{2}\Delta_1(T)-\frac{15}{4}\Delta_0+\frac{25}{4},\end{aligned}\tag{3.3.5}$$

$$\begin{aligned}X_2(TT)&=\Delta_1(T)+\frac{5}{4}(X_1(T)-2\Delta_1(T))\\&=-\frac{3}{2}\Delta_1(T)-\frac{15}{4}\Delta_0+\frac{25}{4}.\end{aligned}\tag{3.3.6}$$

Therefore,

$$\begin{aligned}&\mathbb{E}\ln X_2\\&=\frac{4}{9}\ln\left(6\Delta_1(H)+\frac{15}{4}\Delta_0+\frac{25}{4}\right)+\frac{2}{9}\ln\left(-6\Delta_1(H)+\frac{15}{4}\Delta_0+\frac{25}{4}\right)\\&\quad+\frac{2}{9}\ln\left(\frac{3}{2}\Delta_1(T)-\frac{15}{4}\Delta_0+\frac{25}{4}\right)+\frac{1}{9}\ln\left(-\frac{3}{2}\Delta_1(T)-\frac{15}{4}\Delta_0+\frac{25}{4}\right).\end{aligned}$$

The goal is to maximize this last expression. Toward this end, we compute the partial derivatives

$$\begin{aligned}\frac{\partial}{\partial \Delta_0}\mathbb{E}\ln X_2 &= \frac{4}{9}\cdot\frac{15}{4}\cdot\frac{1}{X_2(HH)}+\frac{2}{9}\cdot\frac{15}{4}\cdot\frac{1}{X_2(HT)}\\ &\quad -\frac{2}{9}\cdot\frac{15}{4}\cdot\frac{1}{X_2(TH)}-\frac{1}{9}\cdot\frac{15}{4}\cdot\frac{1}{X_2(TT)}\\ &= \frac{5}{12}\left(\frac{4}{X_2(HH)}+\frac{2}{X_2(HT)}-\frac{2}{X_2(TH)}-\frac{1}{X_2(TT)}\right),\end{aligned}$$

$$\begin{aligned}\frac{\partial}{\partial \Delta_1(H)}\mathbb{E}\ln X_2 &= \frac{4}{9}\cdot\frac{6}{X_2(HH)}-\frac{2}{9}\cdot\frac{6}{X_2(HT)}\\ &= \frac{4}{3}\left(\frac{2}{X_2(HH)}-\frac{1}{X_2(HT)}\right),\end{aligned}$$

$$\begin{aligned}\frac{\partial}{\partial \Delta_1(T)}\mathbb{E}\ln X_2 &= \frac{2}{9}\cdot\frac{3}{2}\cdot\frac{1}{X_2(TH)}-\frac{1}{9}\cdot\frac{3}{2}\cdot\frac{1}{X_2(TT)}\\ &= \frac{1}{6}\left(\frac{2}{X_2(TH)}-\frac{1}{X_2(TT)}\right).\end{aligned}$$

Setting these derivatives equal to zero, we obtain the three equations

$$\frac{4}{X_2(HH)}+\frac{2}{X_2(HT)}=\frac{2}{X_2(TH)}+\frac{1}{X_2(TT)}, \tag{3.3.7}$$

$$\frac{2}{X_2(HH)}=\frac{1}{X_2(HT)}, \tag{3.3.8}$$

$$\frac{2}{X_2(TH)}=\frac{1}{X_2(TT)}. \tag{3.3.9}$$

We can cross multiply in (3.3.8) and (3.3.9) to obtain

$$X_2(HH)=2X_2(HT), \tag{3.3.10}$$

$$X_2(TH)=2X_2(TT). \tag{3.3.11}$$

Substituting these equations into (3.3.7) and again cross multiplying, we obtain a third equation:

$$X_2(HT)=2X_2(TT). \tag{3.3.12}$$

This gives us the three linear equations (3.3.10)–(3.3.12) in the four unknowns $X_2(HH)$, $X_2(HT)$, $X_2(TH)$, and $X_2(TT)$.

One way to conclude is to recall the formulas (3.3.3)–(3.3.6) for $X_2(HH)$, $X_2(HT)$, $X_2(TH)$, and $X_2(TT)$ in terms of the three unknowns $\Delta_1(H)$, $\Delta_1(T)$, and Δ_0, substitute, and solve the resulting three linear equations in three unknowns. This will lead to the solutions

$$\Delta_0=\frac{5}{9},\quad \Delta_1(H)=\frac{25}{54},\quad \Delta_1(T)=\frac{25}{27}. \tag{3.3.13}$$

We have found the optimal portfolio, but the method we have used is not very pleasant. In particular, as the number of periods increases, the number of variables $\Delta_n(\omega)$ grows exponentially, and in the last step we solved a system of linear equations in these variables.

An alternative way to conclude is to seek a fourth equation involving $X_2(HH)$, $X_2(HT)$, $X_2(TH)$, and $X_2(TT)$ to go with the three equations (3.3.10)–(3.3.12) and then solve these four equations in four unknowns. Such a fourth equation is provided by Corollary 2.4.6 of Chapter 2, which in this context says

$$4 = \frac{16}{25}\left[\frac{1}{4}X_2(HH) + \frac{1}{4}X_2(HT) + \frac{1}{4}X_2(TH) + \frac{1}{4}X_2(TT)\right]. \tag{3.3.14}$$

It is now a straightforward matter to solve (3.3.10)–(3.3.12) and (3.3.14) to obtain

$$X_2(HH) = \frac{100}{9},\ X_2(HT) = \frac{50}{9},\ X_2(TH) = \frac{50}{9},\ X_2(TT) = \frac{25}{9}. \tag{3.3.15}$$

We can then find $\Delta_1(H)$, $\Delta_1(T)$, and Δ_0 by the algorithm of Theorem 1.2.2 of Chapter 1. In particular,

$$\begin{aligned}
\Delta_1(H) &= \frac{X_2(HH) - X_2(HT)}{S_2(HH) - S_2(HT)} &&= \frac{25}{54},\\
\Delta_1(T) &= \frac{X_2(TH) - X_2(TT)}{S_2(TH) - S_2(TT)} &&= \frac{25}{27},\\
X_1(H) &= \frac{4}{5}\left[\frac{1}{2}X_2(HH) + \frac{1}{2}X_2(HT)\right] &&= \frac{20}{3},\\
X_1(T) &= \frac{4}{5}\left[\frac{1}{2}X_2(TH) + \frac{1}{2}X_2(TT)\right] &&= \frac{10}{3}, \qquad (3.3.16)\\
\Delta_0 &= \frac{X_1(H) - X_1(T)}{S_1(H) - S_1(T)} &&= \frac{5}{9}.
\end{aligned}$$

□

The second method of concluding the preceding example used equation (3.3.14), which follows from the fact that the expected discounted value of a portfolio process under the risk-neutral measure is always equal to the initial value X_0 (see Corollary 2.4.6 of Chapter 2). In general,

$$\widetilde{\mathbb{E}}\frac{X_N}{(1+r)^N} = X_0. \tag{3.3.17}$$

This equation introduces the risk-neutral measure to the solution of Problem 3.3.1, even though only the actual probability measure appears in the statement of the problem. This suggests that we might replace Problem 3.3.1 by the following problem.

Problem 3.3.3. Given X_0, find a random variable X_N (without regard to a portfolio process) that maximizes

$$\mathbb{E}U(X_N) \tag{3.3.18}$$

subject to

$$\widetilde{\mathbb{E}}\frac{X_N}{(1+r)^N} = X_0. \tag{3.3.19}$$

Lemma 3.3.4. *Suppose $\Delta_0^*, \Delta_1^*, \ldots, \Delta_{N-1}^*$ is an optimal portfolio process for Problem 3.3.1, and X_N^* is the corresponding optimal wealth random variable at time N. Then X_N^* is optimal for Problem 3.3.3. Conversely, suppose X_N^* is optimal for Problem 3.3.3. Then there is a portfolio process $\Delta_0^*, \Delta_1^*, \ldots, \Delta_{N-1}^*$ that starts with initial wealth X_0 and has value X_N^* at time N, and this portfolio process is optimal for Problem 3.3.1.*

PROOF: Assume first that $\Delta_0^*, \Delta_1^*, \ldots, \Delta_{N-1}^*$ is an optimal portfolio process for Problem 3.3.1, and X_N^* is the corresponding optimal wealth random variable at time N. To show that X_N^* is optimal for Problem 3.3.3, we must show that it satisfies the constraint (3.3.19) and that $\mathbb{E}U(X_N) \le \mathbb{E}U(X_N^*)$ for any other X_N that satisfies this constraint. Because it is generated by a portfolio starting with initial wealth X_0, the random variable X_N^* satisfies (3.3.17), which is (3.3.19). Now let X_N be any other random variable satisfying (3.3.19). We may regard X_N as a derivative security, and according to the risk-neutral pricing formula (2.4.11) of Chapter 2, the time-zero price of this derivative security is X_0 appearing in (3.3.19). In particular, beginning with initial wealth X_0, we may construct a portfolio process $\Delta_1, \Delta_2, \ldots, \Delta_{N-1}$ that replicates X_N (i.e., for which the value of the portfolio process at time N is X_N). (See Theorem 1.2.2 of Chapter 1 for the details.) Since X_N^* is an optimal final portfolio random variable for Problem 3.3.1 and X_N is another final portfolio random variable, we must have $\mathbb{E}U(X_N) \le \mathbb{E}U(X_N^*)$. This shows that X_N^* is optimal for Problem 3.3.3.

For the converse, suppose X_N^* is optimal for Problem 3.3.3. Again using Theorem 1.2.2 of Chapter 1, we may construct a portfolio process $\Delta_0^*, \Delta_1^*, \ldots, \Delta_{N-1}^*$ that begins with initial wealth X_0 and whose value at time N is X_N^*. Let $\Delta_0, \Delta_1, \ldots, \Delta_{N-1}$ be another portfolio process, which, starting with initial wealth X_0, leads to some wealth X_N at time N. To show that X_N^* is optimal in Problem 3.3.1, we must show that

$$\mathbb{E}U(X_N) \le \mathbb{E}U(X_N^*). \tag{3.3.20}$$

But X_N satisfies (3.3.17), which is (3.3.19), and X_N^* is optimal for Problem 3.3.3. This implies (3.3.20) and establishes the optimality of X_N^* in Problem 3.3.1. □

Lemma 3.3.4 separates the optimal investment problem, Problem 3.3.1, into two manageable steps: first, find a random variable X_N that solves Problem 3.3.3; and second, construct the portfolio that starts with X_0 and replicates X_N. The second step uses the algorithm of Theorem 1.2.2 of Chapter 1. It remains only to figure out how to perform the first step. Before giving the general method, we examine Problem 3.3.1 within the context of Example 3.3.2.

Example 3.3.2 (continued) We first compute the Radon-Nikodým derivative of $\widetilde{\mathbb{P}}$ with respect to $\mathbb{P}$:

$$Z(HH) = \frac{\widetilde{\mathbb{P}}(HH)}{\mathbb{P}(HH)} = \frac{9}{16}, \quad Z(HT) = \frac{\widetilde{\mathbb{P}}(HT)}{\mathbb{P}(HT)} = \frac{9}{8},$$
$$Z(TH) = \frac{\widetilde{\mathbb{P}}(TH)}{\mathbb{P}(TH)} = \frac{9}{8}, \quad Z(TT) = \frac{\widetilde{\mathbb{P}}(TT)}{\mathbb{P}(TT)} = \frac{9}{4}.$$

To simplify matters, we use subscripts to denote the different values of the state price density ζ:

$$\zeta_1 = \zeta(HH) = \frac{Z(HH)}{(1+r)^2} = \frac{9}{25},$$
$$\zeta_2 = \zeta(HT) = \frac{Z(HT)}{(1+r)^2} = \frac{18}{25},$$
$$\zeta_3 = \zeta(TH) = \frac{Z(TH)}{(1+r)^2} = \frac{18}{25},$$
$$\zeta_4 = \zeta(TT) = \frac{Z(TT)}{(1+r)^2} = \frac{36}{25}.$$

We also use the notation

$$p_1 = \mathbb{P}(HH) = \frac{4}{9}, \quad p_2 = \mathbb{P}(HT) = \frac{2}{9},$$
$$p_3 = \mathbb{P}(TH) = \frac{2}{9}, \quad p_4 = \mathbb{P}(TT) = \frac{1}{9}.$$

Finally, we denote

$$x_1 = X_2(HH), \quad x_2 = X_2(HT),$$
$$x_3 = X_2(TH), \quad x_4 = X_2(TT).$$

With these notations, Problem 3.3.3 may be written as

Find a vector (x_1, x_2, x_3, x_4) *that maximizes* $\sum_{m=1}^{4} p_m U(x_m)$ *subject to* $\sum_{m=1}^{4} p_m \zeta_m x_m = X_0$.

Filling in the numbers and using the fact that the utility function in question is the logarithm, we rewrite this as

Find a vector (x_1, x_2, x_3, x_4) *that maximizes*

$$\frac{4}{9}\ln x_1 + \frac{2}{9}\ln x_2 + \frac{2}{9}\ln x_3 + \frac{1}{9}\ln x_4$$

subject to

$$\frac{4}{9}\cdot\frac{9}{25}x_1+\frac{2}{9}\cdot\frac{18}{25}x_2+\frac{2}{9}\cdot\frac{18}{25}x_3+\frac{1}{9}\cdot\frac{36}{25}x_4=4. \tag{3.3.21}$$

The Lagrangian for this problem is

$$L=\frac{4}{9}\ln x_1+\frac{2}{9}\ln x_2+\frac{2}{9}\ln x_3+\frac{1}{9}\ln x_4$$
$$-\lambda\left(\frac{4}{9}\cdot\frac{9}{25}x_1+\frac{2}{9}\cdot\frac{18}{25}x_2+\frac{2}{9}\cdot\frac{18}{25}x_3+\frac{1}{9}\frac{36}{25}\cdot x_4-4\right).$$

The Lagrange multiplier equations are

$$\frac{\partial}{\partial x_1}L=\frac{4}{9}\left(\frac{1}{x_1}-\lambda\frac{9}{25}\right)=0,$$
$$\frac{\partial}{\partial x_2}L=\frac{2}{9}\left(\frac{1}{x_2}-\lambda\frac{18}{25}\right)=0,$$
$$\frac{\partial}{\partial x_3}L=\frac{2}{9}\left(\frac{1}{x_3}-\lambda\frac{18}{25}\right)=0,$$
$$\frac{\partial}{\partial x_4}L=\frac{1}{9}\left(\frac{1}{x_4}-\lambda\frac{36}{25}\right)=0,$$

which imply

$$x_1=\frac{25}{9\lambda},\ x_2=\frac{25}{18\lambda},\ x_3=\frac{25}{18\lambda},\ x_4=\frac{25}{36\lambda}.$$

We solve for $\frac{1}{\lambda}$ by substituting these formulas into (3.3.21):

$$\frac{4}{9\lambda}+\frac{2}{9\lambda}+\frac{2}{9\lambda}+\frac{1}{9\lambda}=4,$$

which shows that $\frac{1}{\lambda}=4$. We conclude that the optimal wealth at time two is

$$X_2(HH)=x_1=\frac{100}{9},\ X_2(HT)=x_2=\frac{50}{9},$$
$$X_2(TH)=x_3=\frac{50}{9},\quad X_2(TT)=x_4=\frac{25}{9}.$$

This agrees with formula (3.3.15). We can now compute the optimal portfolio process Δ_0, $\Delta_1(H)$, and $\Delta_1(T)$ as we did following that formula. □

In general, the solution of Problem 3.3.3 follows along the lines of the previous example. It is complicated by the fact that both the actual and risk-neutral probability measures appear in the problem formulation. Consequently, we introduce the Radon-Nikodým derivative Z of $\widetilde{\mathbb{P}}$ with respect to $\mathbb{P}$ to rewrite (3.3.19) without reference to the risk-neutral measure. This constraint becomes

$$\mathbb{E}\frac{Z_N X_N}{(1+r)^N} = X_0. \qquad (3.3.19)'$$

We can take this one step further by recalling the state price density $\zeta = \frac{Z}{(1+r)^N}$, in terms of which (3.3.19) can be written as

$$\mathbb{E}\zeta X_N = X_0. \qquad (3.3.19)''$$

In an N-period model, there are $M = 2^N$ possible coin toss sequences ω. Let us list them, labeling them

$$\omega^1, \omega^2, \ldots, \omega^M.$$

We use superscripts to indicate that ω^m is a full sequence of coin tosses, not the mth coin toss of some sequence. Let us define $\zeta_m = \zeta(\omega^m)$, $p_m = \mathbb{P}(\omega^m)$, and $x_m = X_N(\omega^m)$. Then Problem 3.3.3 can be reformulated as follows.

Problem 3.3.5. Given X_0, find a vector $(x_1, x_2, \ldots, x_M)$ that maximizes

$$\sum_{m=1}^{M} p_m U(x_m)$$

subject to

$$\sum_{m=1}^{M} p_m x_m \zeta_m = X_0.$$

The Lagrangian for Problem 3.3.5 is

$$L = \sum_{m=1}^{M} p_m U(x_m) - \lambda \left(\sum_{m=1}^{M} p_m x_m \zeta_m - X_0 \right),$$

and the Lagrange multiplier equations are

$$\frac{\partial}{\partial x_m} L = p_m U'(x_m) - \lambda p_m \zeta_m = 0, \; m = 1, 2, \ldots, M. \qquad (3.3.22)$$

These equations reduce to

$$U'(x_m) = \lambda \zeta_m, \; m = 1, 2, \ldots, M. \qquad (3.3.23)$$

Recalling how x_m and ζ_m were defined, we rewrite this as

$$U'(X_N) = \frac{\lambda Z}{(1+r)^N}. \qquad (3.3.24)$$

At this point, we need to invert the function U'. Since U is strictly concave everywhere it is finite, its derivative is decreasing and so has an inverse function, which we call I. For example, if $U(x) = \ln x$, then $U'(x) = \frac{1}{x}$. Setting

$y = U'(x) = \frac{1}{x}$, we solve for $x = \frac{1}{y}$, and this determines the inverse function $I(y) = \frac{1}{y}$. After determining this inverse function, whatever it is, we invert (3.3.24) to obtain

$$X_N = I\left(\frac{\lambda Z}{(1+r)^N}\right). \tag{3.3.25}$$

This gives a formula for the optimal X_N in terms of the multiplier λ. We solve for the multiplier λ by substituting X_N into (3.3.19)$'$:

$$\mathbb{E}\left[\frac{Z}{(1+r)^N} I\left(\frac{\lambda Z}{(1+r)^N}\right)\right] = X_0. \tag{3.3.26}$$

After solving this equation for λ, we substitute λ into (3.3.25) to obtain X_N, and then we use the algorithm in Theorem 1.2.2 of Chapter 1 to find the optimal portfolio process $\Delta_0, \Delta_1, \ldots, \Delta_{N-1}$. All these steps were carried out in Example 3.3.2 (continued).

We summarize this discussion with a theorem.

Theorem 3.3.6. *The solution of Problem 3.3.1 can be found by first solving equation (3.3.26) for λ, then computing X_N by (3.3.25), and finally using X_N in the algorithm of Theorem 1.2.2 of Chapter 1 to determine the optimal portfolio process $\Delta_0, \Delta_1, \ldots, \Delta_{N-1}$ and corresponding portfolio value process $X_1, X_2, \ldots, X_N$. The function I appearing in (3.3.26) is the functional inverse of the derivative U' of the utility function U in Problem 3.3.1; i.e., $x = I(y)$ if and only if $y = U'(x)$.*

3.4 Summary

This chapter details the methodology for changing from the actual probability measure to the risk-neutral probability measure in a binomial model and, more generally, the methodology for changing from one probability measure to another in a finite probability model. The key quantity is the *Radon-Nikodým derivative*

$$Z(\omega) = \frac{\widetilde{\mathbb{P}}(\omega)}{\mathbb{P}(\omega)}, \tag{3.1.1}$$

which in a finite probability model is just the quotient of the two probability measures. The Radon-Nikodým derivative is a strictly positive random variable with $\mathbb{E}Z = 1$. The expectations of a random variable Y under the two probability measures are related by the formula

$$\widetilde{\mathbb{E}}Y = \mathbb{E}[ZY]. \tag{3.1.2}$$

In the binomial model with actual probabilities p and q and risk-neutral probabilities $\tilde{p}$ and $\tilde{q}$ for head and tail, respectively, in addition to the *Radon-Nikodým derivative random variable*

$$Z(\omega_1 \dots \omega_N) = \frac{\widetilde{\mathbb{P}}(\omega_1 \dots \omega_N)}{\mathbb{P}(\omega_1 \dots \omega_N)} = \left(\frac{\tilde{p}}{p}\right)^{\#H(\omega_1 \dots \omega_N)} \left(\frac{\tilde{q}}{q}\right)^{\#T(\omega_1 \dots \omega_N)}, \tag{3.1.8}$$

we have a *Radon-Nikodým derivative process*

$$Z_n = \mathbb{E}_n Z, \; n = 0, 1, \dots, N. \tag{3.2.1}$$

This process is also given by the formula

$$Z_n(\omega_1 \dots \omega_n) = \left(\frac{\tilde{p}}{p}\right)^{\#H(\omega_1 \dots \omega_n)} \left(\frac{\tilde{q}}{q}\right)^{\#T(\omega_1 \dots \omega_n)}. \tag{3.2.4}$$

In other words, $Z_n(\omega_1 \dots \omega_n)$ is the ratio of the risk-neutral probability of the partial path of n tosses to the actual probability of the same partial path. When the random variable Y depends only on the first n tosses, where $0 \le n \le N$, equation (3.1.2) takes the simpler form

$$\widetilde{\mathbb{E}} Y = \mathbb{E}[Z_n Y]. \tag{3.2.3}$$

This shows that when Y is determined by the outcome of the first n coin tosses, then we need only consider the ratio of the risk-neutral probability to the actual probability for these n tosses in order to relate $\widetilde{\mathbb{P}}$ and $\mathbb{P}$ expectations of Y.

Conditional expectations under $\widetilde{\mathbb{P}}$ and $\mathbb{P}$ are related as follows. If Y depends only on the first m coin tosses and $0 \le n \le m \le N$, then

$$\widetilde{\mathbb{E}}_n Y = \frac{1}{Z_n} \mathbb{E}_n[Z_m Y]. \tag{3.2.5}$$

When computing the conditional expectation $\widetilde{\mathbb{E}}_n Y$, we imagine we have seen the coin tosses $\omega_1 \dots \omega_n$ and we have assumed that Y does not depend on the tosses $\omega_{m+1} \dots \omega_N$. The coin tosses $\omega_{n+1} \dots \omega_m$, which we have not seen and affect the value of Y, have $\widetilde{\mathbb{P}}$-probability $\tilde{p}^{\#H(\omega_{n+1} \dots \omega_m)} \tilde{q}^{\#T(\omega_{n+1} \dots \omega_m)}$ and $\mathbb{P}$-probability $p^{\#H(\omega_{n+1} \dots \omega_m)} q^{\#T(\omega_{n+1} \dots \omega_m)}$. The ratio of these two probabilities is

$$\frac{Z_m(\omega_1 \dots \omega_m)}{Z_n(\omega_1 \dots \omega_n)} = \left(\frac{\tilde{p}}{p}\right)^{\#H(\omega_{n+1} \dots \omega_m)} \left(\frac{\tilde{q}}{q}\right)^{\#T(\omega_{n+1} \dots \omega_m)},$$

and thus this quotient random variable is used to write the $\widetilde{\mathbb{P}}$-conditional expectation in terms of the $\mathbb{P}$-conditional expectation in (3.2.5). Note in this regard that the right-hand side of (3.2.5) may also be written as $\mathbb{E}_n\left[\frac{Z_m}{Z_n} Y\right]$ since Z_n depends only on the first n coin tosses.

The Radon-Nikodým derivative random variable Z in the binomial model gives rise to the *state price density*

$$\zeta(\omega) = \frac{Z(\omega)}{(1+r)^N}.$$

We may interpret $\zeta(\omega)$ as the value at time zero per unit of actual probability of a derivative security that pays 1 at time N if the coin tossing results in the sequence ω (see the discussion following Definition 3.1.3). In other words, $\zeta(\omega)\mathbb{P}(\omega)$, the so-called *state price* of ω, is the value at time zero of a derivative security that pays 1 at time N if the coin tossing results in the sequence ω.

We can use the state price density to solve the following optimal investment problem.

Problem 3.3.1 (Optimal investment) Given X_0, find an adapted portfolio process $\Delta_0, \Delta_1, \dots, \Delta_{N-1}$ that maximizes

$$\mathbb{E}U(X_N), \tag{3.3.2}$$

subject to the wealth equation

$$X_{n+1} = \Delta_n S_{n+1} + (1+r)(X_n - \Delta_n S_n),\ n = 0, 1, \dots, N-1.$$

If we list the $M = 2^N$ possible coin tosses sequences ω in the N-period binomial model, labeling them $\omega^1, \omega^2, \dots, \omega^M$, and if we define $\zeta_m = \zeta(\omega^m)$, $p_m = \mathbb{P}(\omega^m)$, and $x_m = X_N(\omega^m)$, then Problem 3.3.1 may be reduced to the following problem.

Problem 3.3.5 Given X_0, find a vector $(x_1, x_2, \dots, x_M)$ that maximizes

$$\sum_{m=1}^{M} p_m U(x_m)$$

subject to

$$\sum_{m=1}^{M} p_m x_m \zeta_m = X_0.$$

In Problem 3.3.1, the search is over all portfolio processes, whereas in Problem 3.3.5 the search is over the M variables $x_1, \dots, x_M$. The second problem is simpler and can be solved by the method of Lagrange multipliers. The optimal values of x_m satisfy

$$U'(x_m) = \lambda \zeta_m,\ m = 1, 2, \dots, M, \tag{3.3.23}$$

where λ is the Lagrange multiplier. This leads to the formula

$$X_N = I\left(\frac{\lambda Z}{(1+r)^N}\right), \tag{3.3.25}$$

where I is the inverse of the strictly decreasing function U' and the Lagrange multiplier λ is chosen so that the equation

$$\mathbb{E}\left[\frac{Z}{(1+r)^N} I\left(\frac{\lambda Z}{(1+r)^N}\right)\right] = X_0 \tag{3.3.26}$$

holds. Once the optimal terminal wealth X_N is determined by these equations, we treat it as if it were the payoff of a derivative security and determine the portfolio process $\Delta_0, \Delta_1, \dots, \Delta_{N-1}$ which solves Problem 3.3.1 as the hedge for a short position in this derivative security, using the algorithm of Theorem 1.2.2 of Chapter 1.

3.5 Notes

One might think that the value at time zero of a payoff of 1 at time N if the coin tosses result in "state" ω should be just the expected payoff, which is the (actual rather than risk-neutral) probability of ω. However, this does not account for the risk that the payoff is zero if "state" omega does not occur. The creation of a *state price* to account for risk traces back to Arrow and Debreu [1].

The problem of optimal investment or, more generally, optimal consumption and investment, has been the subject of a great deal of research. The original papers are Hakansson [16] for the discrete-time model and Merton [32], [33], [35] for the continuous-time model. (These papers of Merton are collected in [36].) The solution of the problem via the state price density process is due to Pliska [37] and was further developed by Cox and Huang [7], [8] and Karatzas, Lehoczky, and Shreve [27]. Except for [16], all these models are in continuous time. They have been specialized to the binomial model for this text. A compilation of research along these lines, including treatment of portfolio constraints and the model with different interest rates for borrowing and investment, is provided by Karatzas and Shreve [28].

3.6 Exercises

Exercise 3.1. Under the conditions of Theorem 3.1.1, show the following analogues of properties (i)–(iii) of that theorem:

(i′) $\widetilde{\mathbb{P}}\left(\frac{1}{Z} > 0\right) = 1$;
(ii′) $\widetilde{\mathbb{E}}\frac{1}{Z} = 1$;
(iii′) for any random variable Y,

$$\mathbb{E}Y = \widetilde{\mathbb{E}}\left[\frac{1}{Z} \cdot Y\right].$$

In other words, $\frac{1}{Z}$ facilitates the switch from $\widetilde{\mathbb{E}}$ to $\mathbb{E}$ in the same way Z facilitates the switch from $\mathbb{E}$ to $\widetilde{\mathbb{E}}$.

Exercise 3.2. Let $\mathbb{P}$ be a probability measure on a finite probability space Ω. In this problem, we allow the possibility that $\mathbb{P}(\omega) = 0$ for some values of $\omega \in \Omega$. Let Z be a random variable on Ω with the property that $\mathbb{P}(Z \geq 0) = 1$ and $\mathbb{E}Z = 1$. For $\omega \in \Omega$, define $\widetilde{\mathbb{P}}(\omega) = Z(\omega)\mathbb{P}(\omega)$, and for events $A \subset \Omega$, define $\widetilde{\mathbb{P}}(A) = \sum_{\omega \in A} \widetilde{\mathbb{P}}(\omega)$. Show the following.

(i) $\widetilde{\mathbb{P}}$ is a probability measure; i.e., $\widetilde{\mathbb{P}}(\Omega) = 1$.
(ii) If Y is a random variable, then $\widetilde{\mathbb{E}}Y = \mathbb{E}[ZY]$.
(iii) If A is an event with $\mathbb{P}(A) = 0$, then $\widetilde{\mathbb{P}}(A) = 0$.

(iv) Assume that $\mathbb{P}(Z > 0) = 1$. Show that if A is an event with $\widetilde{\mathbb{P}}(A) = 0$, then $\mathbb{P}(A) = 0$.

When two probability measures agree which events have probability zero (i.e., $\mathbb{P}(A) = 0$ if and only if $\widetilde{\mathbb{P}}(A) = 0$), the measures are said to be *equivalent*. From (iii) and (iv) above, we see that $\mathbb{P}$ and $\widetilde{\mathbb{P}}$ are equivalent under the assumption that $\mathbb{P}(Z > 0) = 1$.

(v) Show that if $\mathbb{P}$ and $\widetilde{\mathbb{P}}$ are equivalent, then they agree which events have probability one (i.e., $\mathbb{P}(A) = 1$ if and only if $\widetilde{\mathbb{P}}(A) = 1$).
(vi) Construct an example in which we have only $\mathbb{P}(Z \geq 0) = 1$ and $\mathbb{P}$ and $\widetilde{\mathbb{P}}$ are not equivalent.

In finance models, the risk-neutral probability measure and actual probability measure must always be equivalent. They agree about what is possible and what is impossible.

Exercise 3.3. Using the stock price model of Figure 3.1.1 and the actual probabilities $p = \frac{2}{3}$, $q = \frac{1}{3}$, define the estimates of S_3 at various times by

$$M_n = \mathbb{E}_n[S_3],\ n = 0, 1, 2, 3.$$

Fill in the values of M_n in a tree like that of Figure 3.1.1. Verify that M_n, $n = 0, 1, 2, 3$, is a martingale.

Exercise 3.4. This problem refers to the model of Example 3.1.2, whose Radon-Nikodým process Z_n appears in Figure 3.2.1.

(i) Compute the state price densities

$$\begin{aligned}
&\zeta_3(HHH),\\
&\zeta_3(HHT) = \zeta_3(HTH) = \zeta_3(THH),\\
&\zeta_3(HTT) = \zeta_3(THT) = \zeta_3(TTH),\\
&\zeta_3(TTT)
\end{aligned}$$

explicitly.
(ii) Use the numbers computed in (i) in formula (3.1.10) to find the time-zero price of the Asian option of Exercise 1.8 of Chapter 1. You should get $v_0(4, 4)$ computed in part (ii) of that exercise.
(iii) Compute also the state price densities $\zeta_2(HT) = \zeta_2(TH)$.
(iv) Use the risk-neutral pricing formula (3.2.6) in the form

$$V_2(HT) = \frac{1}{\zeta_2(HT)} \mathbb{E}_2[\zeta_3 V_3](HT),$$

$$V_2(TH) = \frac{1}{\zeta_2(TH)} \mathbb{E}_2[\zeta_3 V_3](TH)$$

to compute $V_2(HT)$ and $V_2(TH)$. You should get $V_2(HT) = v_2(4, 16)$ and $V_2(TH) = v_2(4, 10)$, where $v_2(s, y)$ was computed in part (ii) of Exercise 1.8 of Chapter 1. Note that $V_2(HT) \neq V_2(TH)$.

Exercise 3.5 (Stochastic volatility, random interest rate). Consider the model of Exercise 2.9 of Chapter 2. Assume that the actual probability measure is

$$\mathbb{P}(HH) = \frac{4}{9}, \ \mathbb{P}(HT) = \frac{2}{9}, \ \mathbb{P}(TH) = \frac{2}{9}, \ \mathbb{P}(TT) = \frac{1}{9}.$$

The risk-neutral measure was computed in Exercise 2.9 of Chapter 2.

(i) Compute the Radon-Nikodým derivative $Z(HH)$, $Z(HT)$, $Z(TH)$, and $Z(TT)$ of $\widetilde{\mathbb{P}}$ with respect to $\mathbb{P}$.
(ii) The Radon-Nikodým derivative process Z_0, Z_1, Z_2 satisfies $Z_2 = Z$. Compute $Z_1(H)$, $Z_1(T)$, and Z_0. Note that $Z_0 = \mathbb{E}Z = 1$.
(iii) The version of the risk-neutral pricing formula (3.2.6) appropriate for this model, which does not use the risk-neutral measure, is

$$\begin{aligned} V_1(H) &= \frac{1+r_0}{Z_1(H)}\mathbb{E}_1\left[\frac{Z_2}{(1+r_0)(1+r_1)}V_2\right](H) \\ &= \frac{1}{Z_1(H)(1+r_1(H))}\mathbb{E}_1[Z_2V_2](H), \\ V_1(T) &= \frac{1+r_0}{Z_1(T)}\mathbb{E}_1\left[\frac{Z_2}{(1+r_0)(1+r_1)}V_2\right](T) \\ &= \frac{1}{Z_1(T)(1+r_1(T))}\mathbb{E}_1[Z_2V_2](T), \\ V_0 &= \mathbb{E}\left[\frac{Z_2}{(1+r_0)(1+r_1)}V_2\right]. \end{aligned}$$

Use this formula to compute $V_1(H)$, $V_1(T)$, and V_0 when $V_2 = (S_2 - 7)^+$. Compare the result with your answers in Exercise 2.6(ii) of Chapter 2.

Exercise 3.6. Consider Problem 3.3.1 in an N-period binomial model with the utility function $U(x) = \ln x$. Show that the optimal wealth process corresponding to the optimal portfolio process is given by $X_n = \frac{X_0}{\zeta_n}$, $n = 0, 1, \ldots, N$, where ζ_n is the state price density process defined in (3.2.7).

Exercise 3.7. Consider Problem 3.3.1 in an N-period binomial model with the utility function $U(x) = \frac{1}{p}x^p$, where $p < 1$, $p \neq 0$. Show that the optimal wealth at time N is

$$X_N = \frac{X_0(1+r)^N Z^{\frac{1}{p-1}}}{\mathbb{E}\left[Z^{\frac{p}{p-1}}\right]},$$

where Z is the Radon-Nikodým derivative of $\widetilde{\mathbb{P}}$ with respect to $\mathbb{P}$.

Exercise 3.8. The Lagrange Multiplier Theorem used in the solution of Problem 3.3.5 has hypotheses that we did not verify in the solution of that problem. In particular, the theorem states that if the gradient of the constraint function, which in this case is the vector $(p_1\zeta_1, \dots, p_m\zeta_m)$, is not the zero vector, then the optimal solution must satisfy the Lagrange multiplier equations (3.3.22). This gradient is not the zero vector, so this hypothesis is satisfied. However, even when this hypothesis is satisfied, the theorem does not guarantee that there is an optimal solution; the solution to the Lagrange multiplier equations may in fact minimize the expected utility. The solution could also be neither a maximizer nor a minimizer. Therefore, in this exercise, we outline a different method for verifying that the random variable X_N given by (3.3.25) maximizes the expected utility.

We begin by changing the notation, calling the random variable given by (3.3.25) X_N^* rather than X_N. In other words,

$$X_N^* = I\left(\frac{\lambda}{(1+r)^N} Z\right), \tag{3.6.1}$$

where λ is the solution of equation (3.3.26). This permits us to use the notation X_N for an arbitrary (not necessarily optimal) random variable satisfying (3.3.19). We must show that

$$\mathbb{E}U(X_N) \leq \mathbb{E}U(X_N^*). \tag{3.6.2}$$

(i) Fix $y > 0$, and show that the function of x given by $U(x) - yx$ is maximized by $y = I(x)$. Conclude that

$$U(x) - yx \leq U(I(y)) - yI(y) \text{ for every } x. \tag{3.6.3}$$

(ii) In (3.6.3), replace the dummy variable x by the random variable X_N and replace the dummy variable y by the random variable $\frac{\lambda Z}{(1+r)^N}$. Take expectations of both sides and use (3.3.19) and (3.3.26) to conclude that (3.6.2) holds.

Exercise 3.9 (Maximizing probability of reaching a goal). (Kulldorf [30], Heath [19])
A wealthy investor provides a small amount of money X_0 for you to use to prove the effectiveness of your investment scheme over the next N periods. You are permitted to invest in the N-period binomial model, subject to the condition that the value of your portfolio is never allowed to be negative. If at time N the value of your portfolio X_N is at least γ, a positive constant specified by the investor, then you will be given a large amount of money to manage for her. Therefore, your problem is the following:
Maximize

$$\mathbb{P}(X_N \geq \gamma),$$

where X_N is generated by a portfolio process beginning with the initial wealth X_0 and where the value X_n of your portfolio satisfies

$$X_n \geq 0, \ n = 1, 2, \dots, N.$$

In the way that Problem 3.3.1 was reformulated as Problem 3.3.3, this problem may be reformulated as
Maximize

$$\mathbb{P}(X_N \geq \gamma)$$

subject to

$$\widetilde{\mathbb{E}} \frac{X_N}{(1+r)^N} = X_0,$$

$$X_n \geq 0, \; n = 1, 2, \ldots, N.$$

(i) Show that if $X_N \geq 0$, then $X_n \geq 0$ for all n.

(ii) Consider the function

$$U(x) = \begin{cases} 0, \text{ if } 0 \leq x < \gamma, \\ 1, \text{ if } x \geq \gamma. \end{cases}$$

Show that for each fixed $y > 0$, we have

$$U(x) - yx \leq U(I(y)) - yI(y) \text{ for every } x \geq 0,$$

where

$$I(y) = \begin{cases} \gamma, \text{ if } 0 < y \leq \frac{1}{\gamma}, \\ 0, \text{ if } y > \frac{1}{\gamma}. \end{cases}$$

(iii) Assume there is a solution λ to the equation

$$\mathbb{E}\left[\frac{Z}{(1+r)^N} I\left(\frac{\lambda Z}{(1+r)^N}\right)\right] = X_0. \tag{3.6.4}$$

Following the argument of Exercise 3.8, show that the optimal X_N is given by

$$X_N^* = I\left(\frac{\lambda Z}{(1+r)^N}\right).$$

(iv) As we did to obtain Problem 3.3.5, let us list the $M = 2^N$ possible coin toss sequences, labeling them $\omega^1, \ldots, \omega^M$, and then define $\zeta_m = \zeta(\omega^m)$, $p_m = \mathbb{P}(\omega^m)$. However, here we list these sequences in ascending order of ζ_m i.e., we label the coin toss sequences so that

$$\zeta_1 \leq \zeta_2 \leq \cdots \leq \zeta_M.$$

Show that the assumption that there is a solution λ to (3.6.4) is equivalent to assuming that for some positive integer K we have $\zeta_K < \zeta_{K+1}$ and

$$\sum_{m=1}^{K} \zeta_m p_m = \frac{X_0}{\gamma}. \tag{3.6.5}$$

(v) Show that X_N^* is given by

$$X_N(\omega^m) = \begin{cases} \gamma, \text{ if } m \leq K, \\ 0, \text{ if } m \geq K+1. \end{cases}$$

4

American Derivative Securities

4.1 Introduction

European option contracts specify an expiration date, and if the option is to be exercised, the exercise must occur on the expiration date. An option whose owner can choose to exercise at any time up to and including the expiration date, or to not exercise at all, is called *American*. Because of this early exercise feature, such an option is at least as valuable as its European counterpart. Sometimes the difference in value is negligible or even zero, and then American and European options are close or exact substitutes. We shall see in this chapter that the early exercise feature for a call on a stock paying no dividend is worthless; American and European calls have the same price. In other cases, most notably put options, the value of this early exercise feature, the so-called *early exercise premium*, can be substantial. An intermediate option between American and European is *Bermudan*, an option that permits early exercise but only on a contractually specified finite set of dates.

Because an American option can be exercised at any time prior to its expiration, it can never be worth less than the payoff associated with immediate exercise. This is called the *intrinsic value* of the option.

In contrast to the case for a European option, whose discounted price process is a martingale under the risk-neutral measure, the discounted price process of an American option is a supermartingale under this measure. The holder of this option may fail to exercise at the optimal exercise date, and in this case the option has a tendency to lose value; hence, the supermartingale property. During any period of time in which it is not optimal to exercise, however, the discounted price process behaves like a martingale.

To price an American option, just as with a European option, we shall imagine selling the option in exchange for some initial capital and then consider how to use this capital to hedge the short position in the option. In this case, we need to be ready to pay off the option at all times prior to the expiration date because we do not know when it will be exercised. We determine when, from our point of view, is the worst time for the owner to exercise the option. From the owner's point of view, this is the *optimal exercise time*, and we shall call it that. We then compute the initial capital we need in order to be hedged against exercise at the optimal exercise time. Finally, we show how to invest this capital so that we are hedged even if the owner exercises at a nonoptimal time. We conclude that the initial price of the option is the capital required to be hedged against optimal exercise.

Section 4.2 sets out the basic American derivative security pricing algorithm when the payoff of the derivative is not path-dependent (i.e., the payoff depends only on the current value of the underlying asset, not on any previous values). In order to develop a complete theory of American derivative securities that includes path-dependent securities, we need the notion of *stopping time*, which is introduced in Section 4.3. Armed with this concept, we work out the general theory of American derivative securities in Section 4.4. One consequence of this general theory is that there is no gain from early exercise of an American call on a nondividend-paying stock. This and related results are developed in Section 4.5.

4.2 Non-Path-Dependent American Derivatives

In this section, we develop a pricing algorithm for American derivative securities when the payoff is not path dependent. We first review the pricing algorithm for European derivative securities when the payoff is not path dependent. In an N-period binomial model with up factor u, down factor d, and interest rate r satisfying the no-arbitrage condition $0 < d < 1 + r < u$, consider a derivative security that pays off $g(S_N)$ at time N for some function g. Because the stock price is Markov, we can write the value V_n of this derivative security at each time n as a function v_n of the stock price at that time i.e., $V_n = v_n(S_n)$, $n = 0, 1, \ldots, N$ (Theorem 2.5.8 of Chapter 2). The risk-neutral pricing formula (see (2.4.12) and its Markov simplification (2.5.2) of Chapter 2) implies that, for $0 \leq n \leq N$, the function v_n is defined by the *European algorithm*:

$$v_N(s) = g(s), \tag{4.2.1}$$

$$v_n(s) = \frac{1}{1+r}\Big[\tilde{p}v_{n+1}(us) + \tilde{q}v_{n+1}(ds)\Big],\ n = N-1, N-2, \ldots, 0, \tag{4.2.2}$$

where $\tilde{p} = \frac{1+r-d}{u-d}$ and $\tilde{q} = \frac{u-1-r}{u-d}$ are the risk-neutral probabilities that the stock goes up and down, respectively. The replicating portfolio (which hedges a short position in the option) is given by (see (1.2.17) of Chapter 1)

$$\Delta_n = \frac{v_{n+1}(uS_n) - v_{n+1}(dS_n)}{(u-d)S_n},\ n = 0, 1, \ldots, N. \tag{4.2.3}$$

Now consider an *American derivative security*. Again, a payoff function g is specified. In any period $n \leq N$, the holder of the derivative security can exercise and receive payment $g(S_n)$. (In this section, the payoff depends only on the current stock price S_n at the time of exercise, not on the stock price path.) Thus, the portfolio that hedges a short position should always have value X_n satisfying

$$X_n \geq g(S_n),\ n = 0, 1, \ldots, N. \tag{4.2.4}$$

The value of the derivative security at each time n is at least as much as the so-called *intrinsic value* $g(S_n)$, and the value of the replicating portfolio at that time must equal the value of the derivative security. We permit the holder of the derivative security to never exercise, and thus the value of the derivative security is always nonnegative, even if g can take negative values.

This suggests that to price an American derivative security, we should replace the European algorithm (4.2.2) by the *American algorithm*:

$$v_N(s) = \max\{g(s), 0\}, \tag{4.2.5}$$

$$v_n(s) = \max\left\{g(s), \frac{1}{1+r}\left[\tilde{p}v_{n+1}(us) + \tilde{q}v_{n+1}(ds)\right]\right\}, \quad n = N-1, N-2, \ldots, 0. \tag{4.2.6}$$

Then $V_n = v_n(S_n)$ would be the price of the derivative security at time n.

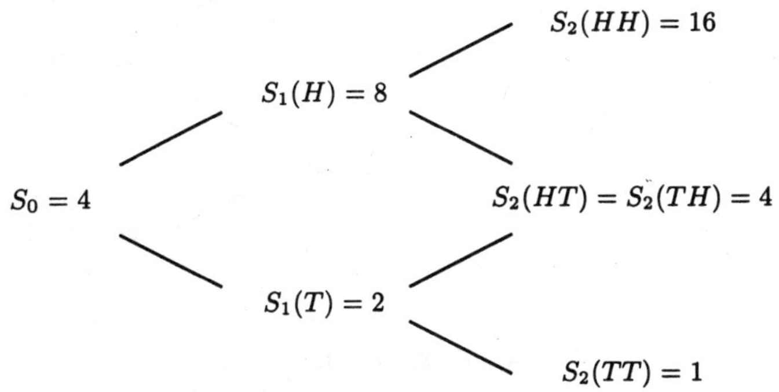

Fig. 4.2.1. A two-period model.

Example 4.2.1. In the two-period model of Figure 4.2.1, let the interest rate be $r = \frac{1}{4}$, so the risk-neutral probabilities are $\tilde{p} = \tilde{q} = \frac{1}{2}$. Consider an American put option, expiring at time two, with strike price 5. In other words, if the owner of the option exercises at time n, she receives $5 - S_n$. We take $g(s) = 5 - s$, and the American algorithm (4.2.5), (4.2.6) becomes

$$v_2(s) = \max\{5 - s, 0\},$$

$$v_n(s) = \max\left\{5 - s, \frac{2}{5}\left[v_{n+1}(2s) + v_{n+1}\left(\frac{s}{2}\right)\right]\right\}, \quad n = 1, 0.$$

In particular,

$$\begin{aligned}
v_2(16) &= 0,\\
v_2(4) &= 1,\\
v_2(1) &= 4,\\
v_1(8) &= \max\left\{(5-8), \frac{2}{5}(0+1)\right\} = \max\{-3, 0.40\} = 0.40,\\
v_1(2) &= \max\left\{(5-2), \frac{2}{5}(1+4)\right\} = \max\{3, 2\} = 3,\\
v_0(4) &= \max\left\{(5-4), \frac{2}{5}(0.40+3)\right\} = \max\{1, 1.36\} = 1.36.
\end{aligned}$$

This algorithm gives a different result than the European algorithm in the

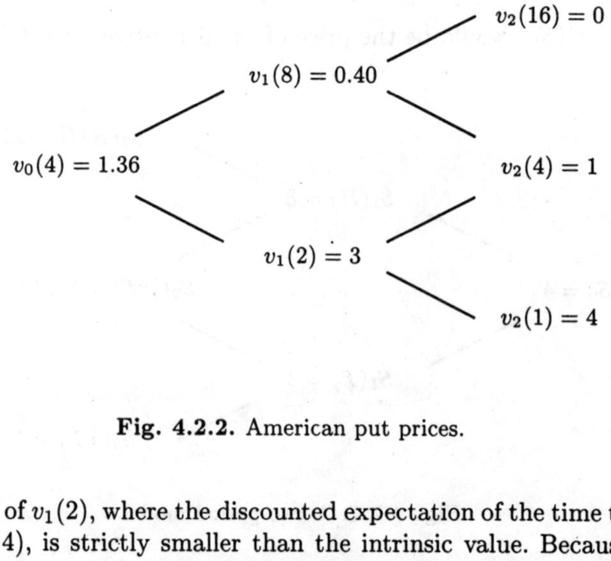

Fig. 4.2.2. American put prices.

computation of $v_1(2)$, where the discounted expectation of the time two option price, $\frac{2}{5}(1+4)$, is strictly smaller than the intrinsic value. Because $v_1(2)$ is strictly greater than the price of a comparable European put, the initial price $v_0(4)$ for the American put is also strictly greater than the initial price of a comparable European put. Figure 4.2.2 shows the American put prices.

Let us now construct the replicating portfolio. We begin with initial capital 1.36 and compute Δ_0 so that the value of the hedging portfolio at time one agrees with the option value. If the first toss results in a head, this requires that

$$\begin{aligned}
0.40 &= v_1(S_1(H))\\
&= S_1(H)\Delta_0 + (1+r)(X_0 - \Delta_0 S_0)
\end{aligned}$$

$$= 8\Delta_0 + \frac{5}{4}(1.36 - 4\Delta_0)$$
$$= 3\Delta_0 + 1.70,$$

which implies that $\Delta_0 = -0.43$. On the other hand, if the first toss results in a tail, we must have

$$\begin{aligned} 3 &= v_1(S_1(T)) \\ &= S_1(T)\Delta_0 + (1+r)(X_0 - \Delta_0 S_0) \\ &= 2\Delta_0 + \frac{5}{4}(1.36 - 4\Delta_0) \\ &= -3\Delta_0 + 1.70, \end{aligned}$$

which also implies that $\Delta_0 = -0.43$. We also could have found this value of Δ_0 by substituting into (4.2.3):

$$\Delta_0 = \frac{v_1(8) - v_1(2)}{8-2} = \frac{0.40 - 3}{8-2} = -0.43.$$

In any case, if we begin with initial capital $X_0 = 1.36$ and take a position of Δ_0 shares of stock at time zero, then at time one we will have $X_1 = V_1 = v_1(S_1)$, regardless of the outcome of the coin toss.

Let us assume that the first coin toss results in a tail. It may be that the owner of the option exercises at time 1, in which case we deliver to her the $3 value of our hedging portfolio and no further hedging is necessary. However, the owner may decline to exercise, in which case the option is still alive and we must continue hedging.

We consider in more detail the case where the owner does not exercise at time 1 after a first toss resulting in tail. We note that next period the option will be worth $v_2(4) = 1$ if the second toss results in head and worth $v_2(1) = 4$ if the second toss results in tail. The risk-neutral pricing formula says that to construct a hedge against these two possibilities, at time 1 we need to have a hedging portfolio valued at

$$\frac{2}{5}\big(v_2(4) + v_2(1)\big) = 2,$$

but we have a hedging portfolio valued at $v_1(2) = 3$. Thus, we may consume $1 and continue the hedge with the remaining $2 value in our portfolio. As this suggests, the option holder has let an optimal exercise time go by.

More specifically, we consume $1 and change our position to $\Delta_1(T)$ shares of stock. If the second coin toss results in head, we want

$$\begin{aligned} 1 &= v_2(S_2(TH)) \\ &= 4\Delta_1(T) + \frac{5}{4}(2 - 2\Delta_1(T)) \\ &= 1.5\Delta_1(T) + 2.50, \end{aligned}$$

and this implies $\Delta_1(T) = -1$. If the second coin toss results in tail, we want

$$\begin{aligned} 4 &= v_2(S_2(TT)) \\ &= \Delta_1(T) + \frac{5}{4}(2 - 2\Delta_1(T)) \\ &= -1.5\Delta_1(T) + 2.50, \end{aligned}$$

and this also implies $\Delta_1(T) = -1$. We could also have gotten this result directly from formula (4.2.3):

$$\Delta_1(T) = \frac{v_2(4) - v_2(1)}{4-1} = \frac{1-4}{4-1} = -1.$$

For the sake of completeness, we consider finally the case where the first toss results in head. At time 1, we will have a portfolio valued at $X_1(H) = 0.40$. We choose

$$\Delta_1(H) = \frac{v_2(16) - v_2(4)}{16-4} = \frac{0-1}{16-4} = -\frac{1}{12}.$$

If the second toss results in head, at time 2 the value of our hedging portfolio is

$$X_2(HH) = 16\Delta_1(H) + \frac{5}{4}(0.40 - 8\Delta_1(H)) = 0 = v_2(16).$$

If the second toss results in tail, at time 2 the value of our hedging portfolio is

$$X_2(HT) = 4\Delta_1(H) + \frac{5}{4}(0.40 - 8\Delta_1(H)) = 1 = v_2(4).$$

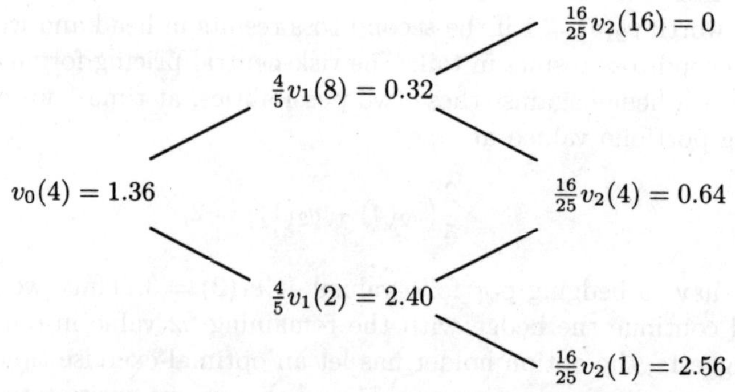

Fig. 4.2.3. Discounted American put prices.

Finally, we consider the discounted American put prices in Figure 4.2.3. These constitute a supermartingale under the risk-neutral probabilities $\tilde{p} =$

$\tilde{q} = \frac{1}{2}$. At each node, the discounted American put price is greater than or equal to the average of the discounted prices at the two subsequent nodes. This price process is not a martingale because the inequality is strict at the time-one node corresponding to a tail on the first toss. □

The following theorem formalizes what we have seen in Example 4.2.1 and justifies the American algorithm (4.2.5), (4.2.6). We shall eventually prove the more general Theorems 4.4.3 and 4.4.4, which cover the case of path dependence as well as path independence, and thus do not pause to prove Theorem 4.2.2 below.

Theorem 4.2.2. (Replication of path-independent American derivatives) *Consider an N-period binomial asset-pricing model with $0 < d < 1+r < u$ and with*

$$\tilde{p} = \frac{1+r-d}{u-d}, \quad \tilde{q} = \frac{u-1-r}{u-d}.$$

Let a payoff function $g(s)$ be given, and define recursively backward in time the sequence of functions $v_N(s), v_{N-1}(s), \ldots, v_0(s)$ by (4.2.5), (4.2.6). Next define

$$\Delta_n = \frac{v_{n+1}(uS_n) - v_{n+1}(dS_n)}{(u-d)S_n}, \tag{4.2.7}$$

$$C_n = v_n(S_n) - \frac{1}{1+r}\left[\tilde{p}v_{n+1}(uS_n) + \tilde{q}v_{n+1}(dS_n)\right], \tag{4.2.8}$$

where n ranges between 0 and $N-1$. We have $C_n \geq 0$ for all n. If we set $X_0 = v_0(S_0)$ and define recursively forward in time the portfolio values $X_1, X_2, \ldots, X_N$ by

$$X_{n+1} = \Delta_n S_{n+1} + (1+r)(X_n - C_n - \Delta_n S_n), \tag{4.2.9}$$

then we will have

$$X_n(\omega_1 \ldots \omega_n) = v_n\big(S_n(\omega_1 \ldots \omega_n)\big) \tag{4.2.10}$$

for all n and all $\omega_1 \ldots \omega_n$. In particular, $X_n \geq g(S_n)$ for all n.

Equation (4.2.9) is the same as the wealth equation (1.2.14) of Chapter 1, except that we have included the possibility of consumption. Theorem 4.2.2 guarantees that we can hedge a short position in the American derivative security with intrinsic value $g(S_n)$ at each time n. In fact, we can do so and perhaps still consume at certain times. The value of our hedging portfolio X_n is always at least as great as the intrinsic value of the derivative security because of (4.2.10) and the fact, guaranteed by (4.2.6), that $v_n(S_n) \geq g(S_n)$. The nonnegativity of C_n also follows from (4.2.6), which implies that

$$v_n(S_n) \geq \frac{1}{1+r}\left[\tilde{p}v_{n+1}(uS_n) + \tilde{q}v_{n+1}(dS_n)\right].$$

4.3 Stopping Times

In general, the time at which an American derivative security should be exercised is random; it depends on the price movements of the underlying asset. We claimed in Example 4.2.1 that if the first coin toss results in tail, then the owner of the American put in that example should exercise at time one. On the other hand, if the first toss results in head, then the owner of the put should not exercise at time one but rather wait for the outcome of the second toss. Indeed, if the first toss results in head, then the stock price is $S_1(H) = 8$ and the put is out of the money. If the second toss results in another head, then $S_2(HH) = 16$, the put is still out of the money, and the owner should let it expire without exercising it. On the other hand, if the first toss is a head and the second toss is a tail, then $S_2(HT) = 4$, the put is in the money at time two, and the owner should exercise. We describe this exercise rule by the following random variable τ:

$$\tau(HH) = \infty, \quad \tau(HT) = 2, \quad \tau(TH) = 1, \quad \tau(TT) = 1, \tag{4.3.1}$$

which is displayed in Figure 4.3.1.

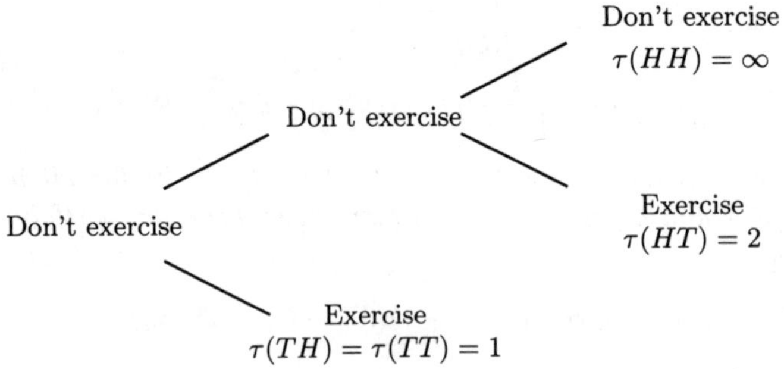

Fig. 4.3.1. Exercise rule τ.

On those paths where τ takes the value ∞, we mean that the option should be allowed to expire without exercise. In Example 4.2.1, there is only one such path, corresponding to HH. In the event the coin tosses result in HT, exercise should be done at time two. In the event the coin tosses result in TH or TT, then exercise should be done at time one. The random variable τ defined on $\Omega = \{HH, HT, TH, TT\}$ by (4.3.1) takes values in the set $\{0, 1, 2, \infty\}$. We can think of it as "stopping" the American put hedging problem by exercising the put, at least on three of the four sample points in Ω. It is a special case of a *stopping time* as defined in Definition 4.3.1 below.

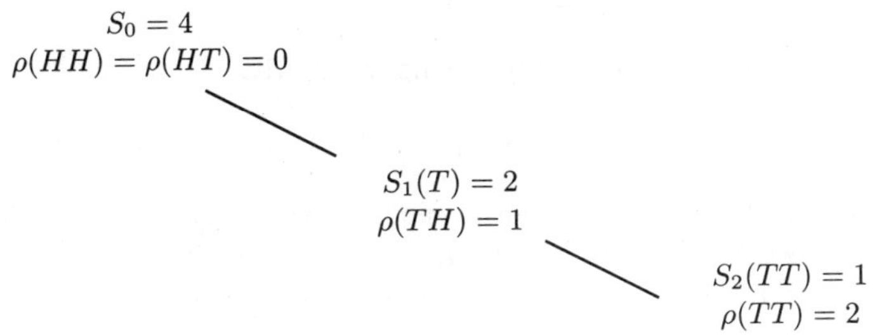

Fig. 4.3.2. Exercise rule ρ.

The owner of the put in this example will regret not exercising the put at time zero if the first coin toss results in a head. In particular, if she had foreknowledge of the coin tossing, she would rather use the exercise rule

$$\rho(HH) = 0, \quad \rho(HT) = 0, \quad \rho(TH) = 1, \quad \rho(TT) = 2, \tag{4.3.2}$$

which is displayed in Figure 4.3.2. If she could use this exercise rule, then regardless of the coin tossing, she would exercise the put in the money. The problem with the exercise rule ρ is that it cannot be implemented without "insider information." It calls for the decision of whether or not to exercise at time zero to be based on the outcome of the first coin toss. If the first coin toss results in a tail, then the decision of whether or not to exercise at time one is based on the outcome of the second coin toss. It is not a stopping time in the sense of Definition 4.3.1 below.

Definition 4.3.1. *In an N-period binomial model, a* stopping time *is a random variable τ that takes values $0, 1, \ldots, N$ or ∞ and satisfies the condition that if $\tau(\omega_1\omega_2 \ldots \omega_n\omega_{n+1} \ldots \omega_N) = n$, then $\tau(\omega_1\omega_2 \ldots \omega_n\omega'_{n+1} \ldots \omega'_N) = n$ for all $\omega'_{n+1} \ldots \omega'_N$.*

The condition in the definition above that if $\tau(\omega_1\omega_2 \ldots \omega_n\omega_{n+1} \ldots \omega_N) = n$ then $\tau(\omega_1\omega_2 \ldots \omega_n\omega'_{n+1} \ldots \omega'_N) = n$ for all $\omega'_{n+1} \ldots \omega'_N$ ensures that stopping is based only on available information. If stopping occurs at time n, then this decision is based only on the first n coin tosses and not on the outcome of any subsequent toss.

Whenever we have a stochastic process and a stopping time, we can define a *stopped process* (see Figure 4.3.3). For example, let Y_n be the process of discounted American put prices in Figure 4.2.3; i.e.,

$$Y_0 = 1.36, \quad Y_1(H) = 0.32, \quad Y_1(T) = 2.40,$$
$$Y_2(HH) = 0, \quad Y_2(HT) = Y_2(TH) = 0.64, \quad Y_2(TT) = 2.56.$$

Let τ be the stopping time of (4.3.1). We define the stopped process $Y_{n\wedge\tau}$ by the formulas below. (The notation $n \wedge \tau$ denotes the minimum of n and τ.) We set

$$Y_{0\wedge\tau} = Y_0 = 1.36$$

because $0 \wedge \tau = 0$ regardless of the coin tossing. Similarly,

$$Y_{1\wedge\tau} = Y_1$$

because $1 \wedge \tau = 1$ regardless of the coin tossing. However, $2 \wedge \tau$ depends on the coin tossing; if we get HH or HT, then $2 \wedge \tau = 2$, but if we get TH or TT, we have $2 \wedge \tau = 1$. Therefore, we have the four cases

$$\begin{aligned} &Y_{2\wedge\tau}(HH) = Y_2(HH) = 0, \;\; Y_{2\wedge\tau}(HT) = Y_2(HT) = 0.64, \\ &Y_{2\wedge\tau}(TH) = Y_1(T) = 2.40, \;\; Y_{2\wedge\tau}(TT) = Y_1(T) = 2.40. \end{aligned}$$

Note in this construction that the process continues on past time 1, even if τ takes the value 1. Time is not stopped. However, the *value* of the process is frozen at time τ. A better terminology might be to call $Y_{n\wedge\tau}$ a *frozen process*, but the term *stopped process* is already in universal use.

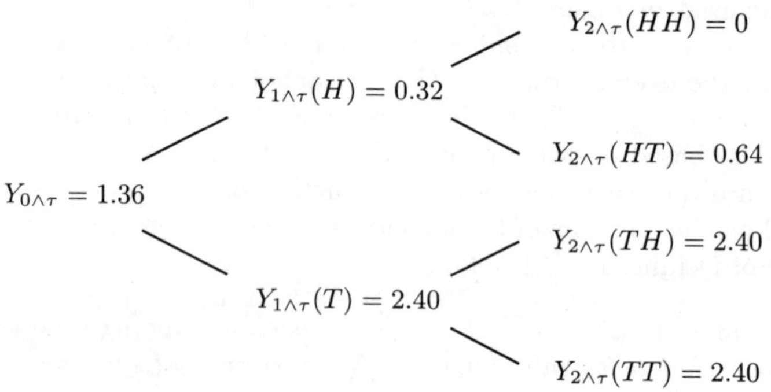

Fig. 4.3.3. A stopped process.

The discounted American put price process $Y_n = \left(\frac{4}{5}\right)^n v_n(S_n)$ in Figure 4.2.3 is a supermartingale but not a martingale under the risk-neutral probabilities $\tilde{p} = \tilde{q} = \frac{1}{2}$ because

$$2.40 = Y_1(T) > \frac{1}{2}Y_2(TH) + \frac{1}{2}Y_2(TT) = \frac{1}{2} \cdot 0.64 + \frac{1}{2} \cdot 2.56 = 1.60.$$

The stopped process $Y_{n\wedge\tau}$ is a martingale. In particular,

$$2.40 = Y_{1\wedge\tau}(T) = \frac{1}{2}Y_{2\wedge\tau}(TH) + \frac{1}{2}Y_{2\wedge\tau}(TT) = \frac{1}{2} \cdot 2.40 + \frac{1}{2} \cdot 2.40.$$

This observation is true generally. Under the risk-neutral probabilities, a discounted American derivative security price process is a supermartingale. However, if this process is stopped at the optimal exercise time, it becomes a martingale. If the owner of the security permits a time to pass in which the supermartingale inequality is strict, she has failed to exercise optimally.

We have just seen by example that we can stop a process that is not a martingale and thereby obtain a martingale. Of course, if we stop a process that is already a martingale, we will obtain a process that may be different but is still a martingale. This trivial observation is one consequence of a general theorem about stopping times called the Optional Sampling Theorem.

Theorem 4.3.2 (Optional sampling—Part I). *A martingale stopped at a stopping time is a martingale. A supermartingale (or submartingale) stopped at a stopping time is a supermartingale (or submartingale, respectively).*

We illustrate the first statement of the theorem with an example. In Figure 4.3.4, we consider the discounted stock price process $M_n = \left(\frac{4}{5}\right)^n S_n$, which is a martingale under the risk-neutral probabilities $\tilde{p} = \tilde{q} = \frac{1}{2}$. At every node, the value is the average of the values at the two subsequent nodes. In Figure 4.3.5, we show the same process stopped by the stopping time τ of (4.3.1). Again, at every node, the value is the average of the values at the two subsequent nodes.

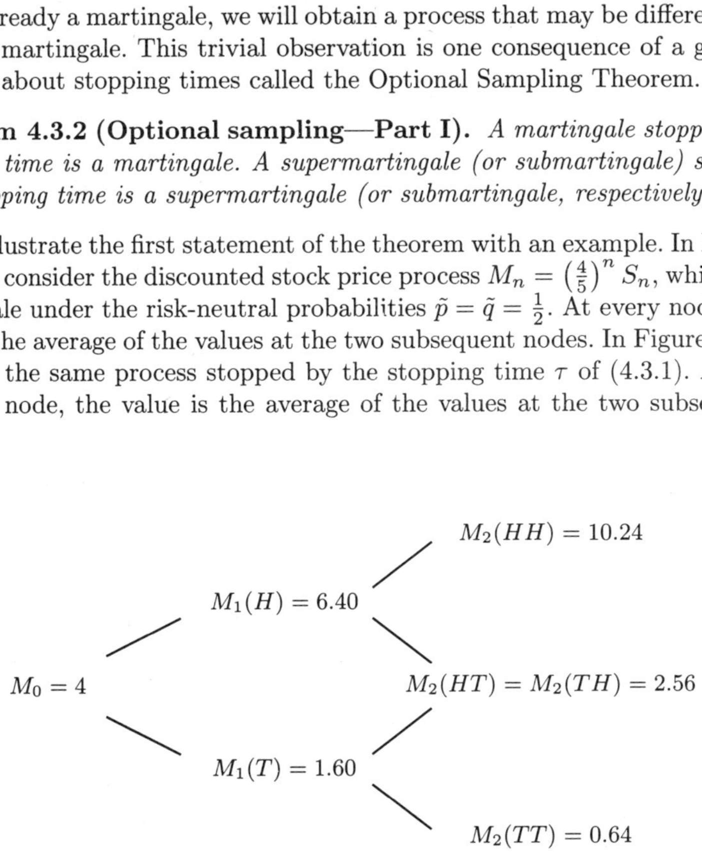

Fig. 4.3.4. Discounted stock price.

Figure 4.3.3 shows a supermartingale stopped at a stopping time, and the resulting process is a martingale, which is of course still a supermartingale. This illustrates the second statement in Theorem 4.3.2.

Finally, we note that if we stop the discounted stock price process of Figure 4.3.4 using the random time ρ of (4.3.2), which is not a stopping time, we destroy the martingale property. Figure 4.3.6 shows the stopped process. The

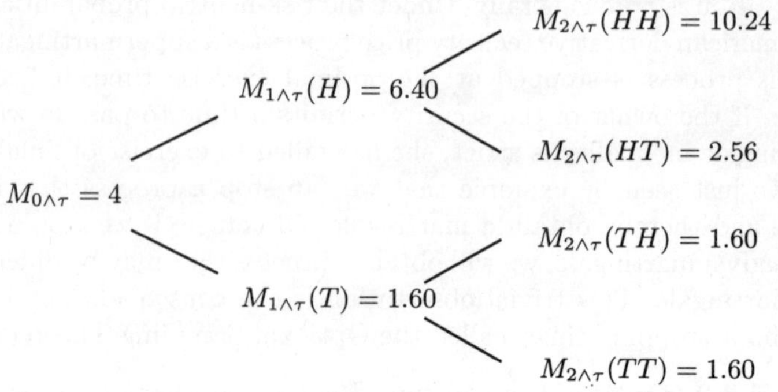

Fig. 4.3.5. Discounted stock price stopped at a stopping time.

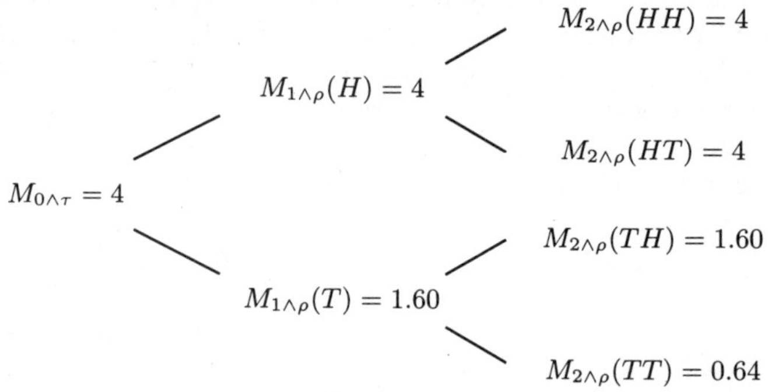

Fig. 4.3.6. Discounted stock price stopped at a nonstopping time.

random time ρ looks ahead and decides to stop if the stock price is about to go up. This introduces a downward bias in the discounted stock price.

A submartingale has a tendency to go up. In particular, if X_n is a submartingale, then $\mathbb{E}X_m \leq \mathbb{E}X_n$ whenever $m \leq n$. This inequality still holds if we replace m by $\tau \wedge n$, where τ is a stopping time.

Theorem 4.3.3 (Optional sampling—Part II). *Let X_n, $n = 0, 1, \ldots, N$ be a submartingale, and let τ be a stopping time. Then $\mathbb{E}X_{n\wedge\tau} \leq \mathbb{E}X_n$. If X_n is a supermartingale, then $\mathbb{E}X_{n\wedge\tau} \geq \mathbb{E}X_n$; if X_n is a martingale, then $\mathbb{E}X_{n\wedge\tau} = \mathbb{E}X_n$.*

The expectation in Theorem 4.3.3 is computed under the probabilities that make X_n be a submartingale (or supermartingale or martingale). In particular, if X_n is a submartingale under the risk-neutral probabilities in a binomial model, then the conclusion of the theorem would be $\widetilde{\mathbb{E}}X_{n\wedge\tau} \leq \widetilde{\mathbb{E}}X_n$.

4.4 General American Derivatives

In this section, we introduce American derivative securities whose intrinsic value is permitted to be path-dependent. We define the price process for such a security and develop its properties. We also show how to hedge a short position in such a derivative security and study the optimal exercise time. All claims made in this section are supported by mathematical proofs.

We work within the context of an N-period binomial model with up factor u, down factor d, and interest rate r satisfying the no-arbitrage condition $0 < d < 1+r < u$. In such a model, we define $\mathcal{S}_n$ to be the set of all stopping times τ that take values in the set $\{n, n+1, \ldots, N, \infty\}$. In particular, the set $\mathcal{S}_0$ contains every stopping time. A stopping time in $\mathcal{S}_N$ can take the value N on some paths, the value ∞ on others, and can take no other value.

Definition 4.4.1. *For each n, $n = 0, 1, \ldots, N$, let G_n be a random variable depending on the first n coin tosses. An* American derivative security with intrinsic value process *G_n is a contract that can be exercised at any time prior to and including time N or not exercised at all, and, if exercised at time n, pays G_n. We define the* price process *V_n for this contract by the* American risk-neutral pricing formula

$$V_n = \max_{\tau \in \mathcal{S}_n} \widetilde{\mathbb{E}}_n \left[\mathbb{I}_{\{\tau \leq N\}} \frac{1}{(1+r)^{\tau-n}} G_\tau \right], \quad n = 0, 1, \ldots, N. \tag{4.4.1}$$

The idea behind (4.4.1) is the following. Suppose the American derivative security is not exercised at times $0, 1, \ldots, n-1$ and we are trying to determine its value at time n. At time n, the owner of the derivative can choose to exercise it immediately or postpone exercise to some later date. The date at which she exercises, if she does exercise, can depend on the path of the stock price up to the exercise time but not beyond it. In other words, the exercise date will be a stopping time τ. Since exercise was not done before time n, this stopping time must be in $\mathcal{S}_n$. Of course, if she never exercises ($\tau = \infty$), then she receives zero payoff. The term $\mathbb{I}_{\{\tau \leq N\}}$ appears in (4.4.1) to tell us that $\mathbb{I}_{\{\tau \leq N\}} \frac{1}{(1+r)^{\tau-n}} G_\tau$ should be replaced by zero on those paths for which $\tau = \infty$. When the owner exercises according to a stopping time $\tau \in \mathcal{S}_n$, the value of the derivative to her at time n is the risk-neutral discounted expectation of its payoff. She should choose τ to make this as large as possible.

One of the immediate consequences of this definition is that

$$V_N = \max\{G_N, 0\}. \tag{4.4.2}$$

To see that, we take $n = N$ in (4.4.1) so that it becomes

$$V_N = \sup_{\tau \in \mathcal{S}_N} \mathbb{I}_{\{\tau \leq N\}} \frac{1}{(1+r)^{\tau-N}} G_\tau.$$

A stopping time in $\mathcal{S}_N$ takes only the values N and ∞, and for such a stopping time

$$\mathbb{I}_{\{\tau\leq N\}}\frac{1}{(1+r)^{\tau-N}}G_\tau = \mathbb{I}_{\{\tau=N\}}G_N.$$

In order to make this as large as possible, we should choose $\tau(\omega_1\ldots\omega_N) = N$ if $G_N(\omega_1\ldots\omega_N) > 0$ and $\tau(\omega_1\ldots\omega_N) = \infty$ if $G_N(\omega_1\ldots\omega_N) < 0$. If $G_N(\omega_1\ldots\omega_N) = 0$, it does not matter whether we take $\tau(\omega_1\ldots\omega_N)$ to be N or ∞. We then have $\mathbb{I}_{\{\tau=N\}}G_N = \max\{G_N, 0\}$, and (4.4.2) is established.

Before working out other consequences of Definition 4.4.1, we rework Example 4.2.1 to verify that this definition is consistent with the American put prices obtained in that example.

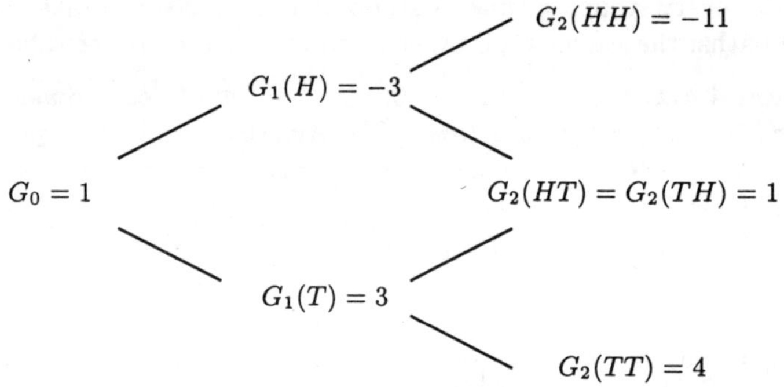

Fig. 4.4.1. Intrinsic value.

Example 4.2.1 continued For the American put with strike 5 on the stock given in Figure 4.2.1, the intrinsic value is given in Figure 4.4.1. In this example, $N = 2$ and (4.4.2) becomes

$$V_2(HH) = 0,\ V_2(HT) = V_2(TH) = 1,\ V_2(TT) = 4.$$

We next apply (4.4.1) with $n = 1$, first considering the case of head on the first toss. Then

$$V_1(H) = \max_{\tau\in\mathcal{S}_1}\widetilde{\mathbb{E}}_1\left[\mathbb{I}_{\{\tau\leq 2\}}\left(\frac{4}{5}\right)^{\tau-1}G_\tau\right](H). \tag{4.4.3}$$

To make the conditional expectation on the right-hand side of (4.4.3) as large as possible, we should take $\tau(HH) = \infty$ (do not exercise in the case of HH) and take $\tau(HT) = 2$ (exercise at time one in the case of HT). The decision of whether or not to exercise at time two is based on the information available at time two, so this does not violate the property required of stopping times. This exercise policy makes (4.4.3) equal to

$$V_1(H) = \frac{1}{2}\cdot 0 + \frac{1}{2}\cdot\left(\frac{4}{5}\right)^{2-1}G_2(HT) = 0.40.$$

We next apply (4.4.1) with $n = 1$ when the first toss results in a tail. In this case,

$$V_1(T) = \max_{\tau \in \mathcal{S}_1} \widetilde{\mathbb{E}}_1 \left[\mathbb{I}_{\{\tau \leq 2\}} \left(\frac{4}{5}\right)^{\tau-1} G_\tau \right] (T). \quad (4.4.4)$$

To make this conditional expectation as large as possible, knowing that the first toss results in a tail, we must consider two possibilities: exercise at time one or exercise at time two. It is clear that we want to exercise at one of these two times because the option is in the money regardless of the second coin toss. If we take $\tau(TH) = \tau(TT) = 1$, then

$$\widetilde{\mathbb{E}}_1 \left[\mathbb{I}_{\{\tau \leq 2\}} \left(\frac{4}{5}\right)^{\tau-1} G_\tau \right] (T) = G_1(T) = 3.$$

If we take $\tau(TH) = \tau(TT) = 2$, then

$$\widetilde{\mathbb{E}}_1 \left[\mathbb{I}_{\{\tau \leq 2\}} \left(\frac{4}{5}\right)^{\tau-1} G_\tau \right] (T) = \frac{4}{5} \left(\frac{1}{2} \cdot 1 + \frac{1}{2} \cdot 4 \right) = 2.$$

We cannot choose $\tau(TH) = 1$ and $\tau(TT) = 2$ because that would violate the property of stopping times. We see then that (4.4.4) yields $V_1(T) = 3$.

Finally, when $n = 0$, we have

$$V_0 = \max_{\tau \in \mathcal{S}_0} \widetilde{\mathbb{E}} \left[\mathbb{I}_{\{\tau \leq 2\}} \left(\frac{4}{5}\right)^{\tau} G_\tau \right]. \quad (4.4.5)$$

There are many stopping times to consider (see Exercise 4.5), but a moment's reflection shows that the one that makes $\widetilde{\mathbb{E}} \left[\mathbb{I}_{\{\tau \leq 2\}} \left(\frac{4}{5}\right)^{\tau} G_\tau \right]$ as large as possible is

$$\tau(HH) = \infty, \ \tau(HT) = 2, \ \tau(TH) = \tau(TT) = 1. \quad (4.4.6)$$

With this stopping time, (4.4.5) becomes

$$V_0 = \frac{1}{4} \cdot 0 + \frac{1}{4} \left(\frac{4}{5}\right)^2 G_2(HT) + \frac{1}{2} \cdot \frac{4}{5} \cdot G_1(T) = \frac{1}{4} \cdot \frac{16}{25} \cdot 1 + \frac{1}{2} \cdot \frac{4}{5} \cdot 3 = 1.36. \quad (4.4.7)$$

We record the option prices in Figure 4.4.2. These agree with the prices in Figure 4.2.2, the only difference being that in Figure 4.2.2 these prices are recorded as functions v_n of the underlying stock prices and here they are recorded as random variables (i.e., functions of the coin tosses). The prices in the two figures are related by the formula $V_n = v_n(S_n)$. □

We now develop the properties of the American derivative security price process of Definition 4.4.1. These properties justify calling V_n defined by (4.4.1) the *price* of the derivative security.

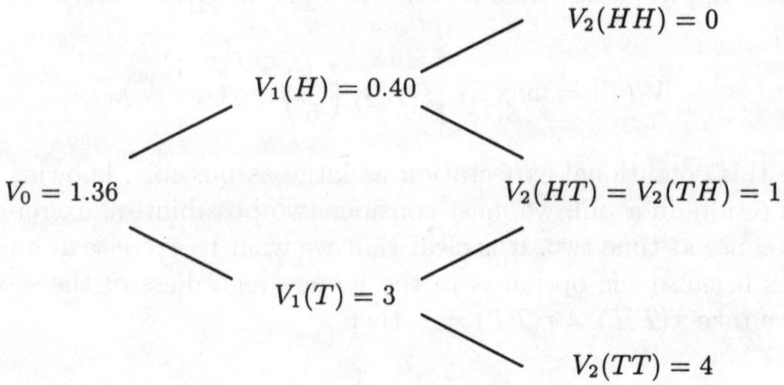

Fig. 4.4.2. American put prices.

Theorem 4.4.2. *The American derivative security price process given by Definition 4.4.1 has the following properties:*

(i) $V_n \geq \max\{G_n, 0\}$ *for all n;*
(ii) the discounted process $\frac{1}{(1+r)^n} V_n$ *is a supermartingale;*
(iii) if Y_n *is another process satisfying* $Y_n \geq \max\{G_n, 0\}$ *for all n and for which* $\frac{1}{(1+r)^n} Y_n$ *is a supermartingale, then* $Y_n \geq V_n$ *for all n.*

We summarize property (iii) by saying that V_n *is the* smallest *process satisfying (i) and (ii).*

We shall see that property (ii) in Theorem 4.4.2 guarantees that an agent beginning with initial capital V_0 can construct a hedging portfolio whose value at each time n is V_n. Property (i) guarantees that if an agent does this, he has hedged a short position in the derivative security; no matter when it is exercised, the agent's hedging portfolio value is sufficient to pay off the derivative security. Thus, (i) and (ii) guarantee that the derivative security price is acceptable to the seller. Condition (iii) says that the price is no higher than necessary in order to be acceptable to the seller. This condition ensures that the price is fair for the buyer.

PROOF: We first establish (i). Let n be given, and consider the stopping time $\widehat{\tau}$ in $\mathcal{S}_n$ that takes the value n, regardless of the coin tossing. Then

$$\widetilde{\mathbb{E}}_n\left[\mathbb{I}_{\{\widehat{\tau}\leq N\}}\frac{1}{(1+r)^{\widehat{\tau}-n}}G_{\widehat{\tau}}\right] = G_n.$$

Since V_n is the largest possible value we can obtain for $\widetilde{\mathbb{E}}_n\left[\mathbb{I}_{\{\tau\leq N\}}\frac{1}{(1+r)^{\tau-n}}G_\tau\right]$ when we consider all stopping times $\tau \in \mathcal{S}_n$, we must have $V_n \geq G_n$. On the other hand, if we take $\overline{\tau}$ to be the stopping time in $\mathcal{S}_n$ that takes the value ∞, regardless of the coin tossing, then

$$\widetilde{\mathbb{E}}_n \left[\mathbb{I}_{\{\overline{\tau} \le N\}} \frac{1}{(1+r)^{\overline{\tau}-n}} G_{\overline{\tau}} \right] = 0.$$

Again, V_n is the maximum of expressions of this type, and hence $V_n \ge 0$. We conclude that (i) holds.

We next prove (ii). Let n be given, and suppose τ^* attains the maximum in the definition of V_{n+1}; i.e., $\tau^* \in \mathcal{S}_{n+1}$ and

$$V_{n+1} = \widetilde{\mathbb{E}}_{n+1} \left[\mathbb{I}_{\{\tau^* \le N\}} \frac{1}{(1+r)^{\tau^*-n-1}} G_{\tau^*} \right]. \tag{4.4.8}$$

But $\tau^* \in \mathcal{S}_n$ also, which together with iterated conditioning implies

$$\begin{aligned} V_n &\ge \widetilde{\mathbb{E}}_n \left[\mathbb{I}_{\{\tau^* \le N\}} \frac{1}{(1+r)^{\tau^*-n}} G_{\tau^*} \right] \\ &= \widetilde{\mathbb{E}}_n \left[\widetilde{\mathbb{E}}_{n+1} \left[\mathbb{I}_{\{\tau^* \le N\}} \frac{1}{(1+r)^{\tau^*-n}} G_{\tau^*} \right] \right] \\ &= \widetilde{\mathbb{E}}_n \left[\frac{1}{1+r} \widetilde{\mathbb{E}}_{n+1} \left[\mathbb{I}_{\{\tau^* \le N\}} \frac{1}{(1+r)^{\tau^*-n-1}} G_{\tau^*} \right] \right] \\ &= \widetilde{\mathbb{E}}_n \left[\frac{1}{1+r} V_{n+1} \right]. \end{aligned} \tag{4.4.9}$$

Dividing both sides by $(1+r)^n$, we obtain the supermartingale property for the discounted price process:

$$\frac{1}{(1+r)^n} V_n \ge \widetilde{\mathbb{E}}_n \left[\frac{1}{(1+r)^{n+1}} V_{n+1} \right].$$

Finally, we prove (iii). Let Y_n be another process satisfying conditions (i) and (ii). Let $n \le N$ be given and let τ be a stopping time in $\mathcal{S}_n$. Because $Y_k \ge \max\{G_k, 0\}$ for all k, we have

$$\begin{aligned} \mathbb{I}_{\{\tau \le N\}} G_\tau &\le \mathbb{I}_{\{\tau \le N\}} \max\{G_\tau, 0\} \\ &\le \mathbb{I}_{\{\tau \le N\}} \max\{G_{N \wedge \tau}, 0\} + \mathbb{I}_{\{\tau = \infty\}} \max\{G_{N \wedge \tau}, 0\} \\ &= \max\{G_{N \wedge \tau}, 0\} \\ &\le Y_{N \wedge \tau}. \end{aligned}$$

We next use the Optional Sampling Theorem 4.3.2 and the supermartingale property for $\frac{1}{(1+r)^k} Y_k$ to write

$$\begin{aligned} \widetilde{\mathbb{E}}_n \left[\mathbb{I}_{\{\tau \le N\}} \frac{1}{(1+r)^\tau} G_\tau \right] &= \widetilde{\mathbb{E}}_n \left[\mathbb{I}_{\{\tau \le N\}} \frac{1}{(1+r)^{N \wedge \tau}} G_\tau \right] \\ &\le \widetilde{\mathbb{E}}_n \left[\frac{1}{(1+r)^{N \wedge \tau}} Y_{N \wedge \tau} \right] \\ &\le \frac{1}{(1+r)^{n \wedge \tau}} Y_{n \wedge \tau} \\ &= \frac{1}{(1+r)^n} Y_n, \end{aligned}$$

the equality at the end being a consequence of the fact that $\tau \in \mathcal{S}_n$ is greater than or equal to n on every path. Multiplying by $(1+r)^n$, we obtain

$$\widetilde{\mathbb{E}}_n \left[\mathbb{I}_{\{\tau \le N\}} \frac{1}{(1+r)^{\tau - n}} G_\tau \right] \le Y_n.$$

Since V_n is the maximum value we can obtain for $\widetilde{\mathbb{E}}_n \left[\mathbb{I}_{\{\tau \le N\}} \frac{1}{(1+r)^{\tau-n}} G_\tau \right]$ as τ ranges over $\mathcal{S}_n$, and all these values are less than or equal to Y_n, we must have $V_n \le Y_n$. □

We now generalize the American pricing algorithm given by (4.2.5) and (4.2.6) to path-dependent securities

Theorem 4.4.3. *We have the following American pricing algorithm for the path-dependent derivative security price process given by Definition 4.4.1:*

$$V_N(\omega_1 \dots \omega_N) = \max\{G_N(\omega_1 \dots \omega_N), 0\}, \tag{4.4.10}$$

$$V_n(\omega_1 \dots \omega_n) = \max \left\{ G_n(\omega_1 \dots \omega_n), \frac{1}{1+r} \Big[\tilde{p} V_{n+1}(\omega_1 \dots \omega_n H) + \tilde{q} V_{n+1}(\omega_1 \dots \omega_n T) \Big] \right\}, \tag{4.4.11}$$

for $n = N-1, \dots, 0$.

Proof: We shall prove that V_n defined recursively by (4.4.10) and (4.4.11) satisfies properties (i) and (ii) of Theorem 4.4.2 and is the smallest process with these properties. According to Theorem 4.4.2, the process V_n given by (4.4.1) is the smallest process with these properties, and hence the algorithm (4.4.10), (4.4.11) must generate the same process as formula (4.4.1).

We first establish property (i) of Theorem 4.4.2. It is clear that V_N defined by (4.4.10) satisfies property (i) with $n = N$. We proceed by induction backward in time. Suppose that for some n between 0 and $N-1$ we have $V_{n+1} \ge \max\{G_{n+1}, 0\}$. Then, from (4.4.11), we see that

$$V_n(\omega_1 \dots \omega_n) \ge \max\{G_n(\omega_1 \dots \omega_n), 0\}.$$

This completes the induction step and shows that V_n defined recursively by (4.4.10), (4.4.11) satisfies property (i) of Theorem 4.4.2.

We next verify that $\frac{1}{(1+r)^n} V_n$ is a supermartingale. From (4.4.11), we see immediately that

$$\begin{aligned} V_n(\omega_1 \dots \omega_n) &\ge \frac{1}{1+r} \Big[\tilde{p} V_{n+1}(\omega_1 \dots \omega_n H) + \tilde{q} V_{n+1}(\omega_1 \dots \omega_n T) \Big] \\ &= \widetilde{\mathbb{E}}_n \left[\frac{1}{1+r} V_{n+1} \right] (\omega_1 \dots \omega_n). \end{aligned} \tag{4.4.12}$$

Multiplying both sides by $\frac{1}{(1+r)^n}$, we obtain the desired supermartingale property.

Finally, we must show that V_n defined by (4.4.10), (4.4.11) and satisfying (i) and (ii) of Theorem 4.4.2 is the smallest process satisfying (i) and (ii). It is immediately clear from (4.4.10) that V_N is the smallest random variable satisfying $V_N \geq \max\{G_N, 0\}$. We proceed by induction backward in time. Suppose that, for some n between 0 and $N-1$, V_{n+1} is as small as possible. The supermartingale property (ii) implies that V_n must satisfy (4.4.12). In order to satisfy property (i), V_n must be greater than or equal to G_n. Therefore, properties (i) and (ii) of Theorem 4.4.2 imply

$$V_n(\omega_1 \dots \omega_n) \geq \max\Bigg\{G_n(\omega_1 \dots \omega_n), \frac{1}{1+r}\Big[\tilde{p}V_{n+1}(\omega_1 \dots \omega_n H) + \tilde{q}V_{n+1}(\omega_1 \dots \omega_n T)\Big]\Bigg\} \tag{4.4.13}$$

for $n = N-1, \dots, 0$. But (4.4.11) defines $V_n(\omega_1 \dots \omega_n)$ to be equal to the right-hand side of (4.4.13), which means $V_n(\omega_1 \dots \omega_n)$ is as small as possible. □

In order to justify Definition 4.4.1 for American derivative security prices, we must show that a short position can be hedged using these prices. This requires a generalization of Theorem 4.2.2 to the path-dependent case.

Theorem 4.4.4 (Replication of path-dependent American derivatives). *Consider an N-period binomial asset-pricing model with $0 < d < 1+r < u$ and with*

$$\tilde{p} = \frac{1+r-d}{u-d}, \quad \tilde{q} = \frac{u-1-r}{u-d}.$$

For each n, $n = 0, 1, \dots, N$, let G_n be a random variable depending on the first n coin tosses. With V_n, $n = 0, 1, \dots, N$, given by Definition 4.4.1, we define

$$\Delta_n(\omega_1 \dots \omega_n) = \frac{V_{n+1}(\omega_1 \dots \omega_n H) - V_{n+1}(\omega_1 \dots \omega_n T)}{S_{n+1}(\omega_1 \dots \omega_n H) - S_{n+1}(\omega_1 \dots \omega_n T)}, \tag{4.4.14}$$

$$C_n(\omega_1 \dots \omega_n) = V_n(\omega_1 \dots \omega_n) - \frac{1}{1+r}\Big[\tilde{p}V_{n+1}(\omega_1 \dots \omega_n H) + \tilde{q}V_{n+1}(\omega_1 \dots \omega_n T)\Big], \tag{4.4.15}$$

where n ranges between 0 and $N-1$. We have $C_n \geq 0$ for all n. If we set $X_0 = V_0$ and define recursively forward in time the portfolio values $X_1, X_2, \dots, X_N$ by

$$X_{n+1} = \Delta_n S_{n+1} + (1+r)(X_n - C_n - \Delta_n S_n), \tag{4.4.16}$$

then we have

$$X_n(\omega_1 \dots \omega_n) = V_n(\omega_1 \dots \omega_n) \tag{4.4.17}$$

for all n and all $\omega_1 \dots \omega_n$. In particular, $X_n \geq G_n$ for all n.

PROOF: The nonnegativity of C_n is a consequence of property (ii) of Theorem 4.4.2 or, equivalently, (4.4.12).

To prove (4.4.17), we proceed by induction on n. This part of the proof is the same as the proof of Theorem 2.4.8. The induction hypothesis is that $X_n(\omega_1 \dots \omega_n) = V_n(\omega_1 \dots \omega_n)$ for some $n \in \{0, 1, \dots, N-1\}$ and all $\omega_1 \dots \omega_n$. We need to show that

$$X_{n+1}(\omega_1 \dots \omega_n H) = V_{n+1}(\omega_1 \dots \omega_n H), \tag{4.4.18}$$

$$X_{n+1}(\omega_1 \dots \omega_n T) = V_{n+1}(\omega_1 \dots \omega_n T). \tag{4.4.19}$$

We prove (4.4.18); the proof of (4.4.19) is analogous.

Note first that

$$\begin{aligned} &V_n(\omega_1 \dots \omega_n) - C_n(\omega_1 \dots \omega_n) \\ &= \frac{1}{1+r}\Big[\tilde{p}V_{n+1}(\omega_1 \dots \omega_n H) + \tilde{q}V_{n+1}(\omega_1 \dots \omega_n T)\Big]. \end{aligned}$$

Since $\omega_1 \dots \omega_n$ will be fixed for the rest of the proof, we will suppress these symbols. For example, the last equation will be written simply as

$$V_n - C_n = \frac{1}{1+r}\Big[\tilde{p}V_{n+1}(H) + \tilde{q}V_{n+1}(T)\Big].$$

We compute

$$\begin{aligned} X_{n+1}(H) &= \Delta_n S_{n+1}(H) + (1+r)(X_n - C_n - \Delta_n S_n) \\ &= \frac{V_{n+1}(H) - V_{n+1}(T)}{S_{n+1}(H) - S_{n+1}(T)}\big(S_{n+1}(H) - (1+r)S_n\big) \\ &\qquad +(1+r)(V_n - C_n) \\ &= \frac{V_{n+1}(H) - V_{n+1}(T)}{(u-d)S_n}\big(uS_n - (1+r)S_n\big) \\ &\qquad +\tilde{p}V_{n+1}(H) + \tilde{q}V_{n+1}(T) \\ &= \big(V_{n+1}(H) - V_{n+1}(T)\big)\frac{u-1-r}{u-d} + \tilde{p}V_{n+1}(H) + \tilde{q}V_{n+1}(T) \\ &= \big(V_{n+1}(H) - V_{n+1}(T)\big)\tilde{q} + \tilde{p}V_{n+1}(H) + \tilde{q}V_{n+1}(T) \\ &= (\tilde{p} + \tilde{q})V_{n+1}(H) \;=\; V_{n+1}(H). \end{aligned}$$

This is (4.4.18).

The final claim of the theorem, that $X_n \geq G_n$ for all n, follows from (4.4.17) and property (i) of Theorem 4.4.2. □

Theorem 4.4.4 shows that the American derivative security price given by (4.4.1) is acceptable to the seller because he can construct a hedge for the short position. We next argue that it is also acceptable to the buyer. Let us fix n, imagine we have gotten to time n without the derivative security being exercised, and denote by $\tau^* \in \mathcal{S}_n$ the stopping time that attains the maximum in (4.4.1), so that

$$V_n = \widetilde{\mathbb{E}}_n \left[\mathbb{I}_{\{\tau^* \leq N\}} \frac{1}{(1+r)^{\tau^* - n}} G_{\tau^*} \right]. \tag{4.4.20}$$

For $k = n, n+1, \ldots, N$, define

$$C_k = \mathbb{I}_{\{\tau^* = k\}} G_k.$$

If the owner of the derivative security exercises it according to the stopping time τ^*, then she will receive the cash flows $C_n, C_{n+1}, \ldots, C_N$ at times $n, n+1, \ldots, N$, respectively. Actually, at most one of these C_k values is non-zero. If the option is exercised at or before the expiration time N, then the C_k corresponding to the exercise time is the only nonzero payment among them. However, on different paths, this payment comes at different times. In any case, (4.4.20) becomes

$$V_n = \widetilde{\mathbb{E}}_n \left[\sum_{k=n}^{N} \mathbb{I}_{\{\tau^* = k\}} \frac{1}{(1+r)^{k-n}} G_k \right] = \widetilde{\mathbb{E}}_n \left[\sum_{k=n}^{N} \frac{C_k}{(1+r)^{k-n}} \right].$$

We saw in Theorem 2.4.8 of Chapter 2 that this is just the value at time n of the cash flows $C_n, C_{n+1}, \ldots, C_N$, received at times $n, n+1, \ldots, N$, respectively. Once the option holder decides on the exercise strategy τ^*, this is exactly the contract she holds. Thus, the American derivative security price V_n is acceptable to her.

It remains to provide a method for the American derivative security owner to choose an optimal exercise time. We shall consider this problem with $n = 0$ (i.e., seek a stopping time $\tau^* \in \mathcal{S}_0$ that achieves the maximum in (4.4.1) when $n = 0$).

Theorem 4.4.5 (Optimal exercise). *The stopping time*

$$\tau^* = \min\{n; V_n = G_n\} \tag{4.4.21}$$

maximizes the right-hand side of (4.4.1) when $n = 0$; i.e.,

$$V_0 = \widetilde{\mathbb{E}} \left[\mathbb{I}_{\{\tau^* \leq N\}} \frac{1}{(1+r)^{\tau^*}} G_{\tau^*} \right]. \tag{4.4.22}$$

The value of an American derivative security is always greater than or equal to its intrinsic value. The stopping time τ^* of (4.4.21) is the first time these two are equal. In may be that they are never equal. For example, the value of an American put is always greater than or equal to zero, but the put can always be out of the money (i.e., with negative intrinsic value). In this case, the minimum in (4.4.21) is over the empty set (the set of integers n for which $V_n = G_n$ is the empty set), and we follow the mathematical convention that the minimum over the empty set is ∞. For us, $\tau^* = \infty$ is synonymous with the derivative security expiring unexercised.

PROOF OF THEOREM 4.4.5: We first observe that the stopped process

$$\frac{1}{(1+r)^{n\wedge\tau^*}} V_{n\wedge\tau^*} \tag{4.4.23}$$

is a martingale under the risk-neutral probability measure. This is a consequence of (4.4.11). Indeed, if the first n coin tosses result in $\omega_1 \dots \omega_n$ and along this path $\tau^* \geq n+1$, then we know that $V_n(\omega_1 \dots \omega_n) > G_n(\omega_1 \dots \omega_n)$ and (4.4.11) implies

$$\begin{aligned}
&V_{n\wedge\tau^*}(\omega_1 \dots \omega_n)\\
&= V_n(\omega_1 \dots \omega_n)\\
&= \frac{1}{1+r}\Big[\tilde{p}V_{n+1}(\omega_1 \dots \omega_n H) + \tilde{q}V_{n+1}(\omega_1 \dots \omega_n T)\Big]\\
&= \frac{1}{1+r}\Big[\tilde{p}V_{(n+1)\wedge\tau^*}(\omega_1 \dots \omega_n H) + \tilde{q}V_{(n+1)\wedge\tau^*}(\omega_1 \dots \omega_n T)\Big].
\end{aligned}$$

This is the martingale property for the process (4.4.23). On the other hand, if along the path $\omega_1 \dots \omega_n$ we have $\tau^* \leq n$, then

$$\begin{aligned}
&V_{n\wedge\tau^*}(\omega_1 \dots \omega_n)\\
&= V_{\tau^*}(\omega_1 \dots \omega_{\tau^*})\\
&= \tilde{p}V_{\tau^*}(\omega_1 \dots \omega_{\tau^*}) + \tilde{q}V_{\tau^*}(\omega_1 \dots \omega_{\tau^*})\\
&= \tilde{p}V_{(n+1)\wedge\tau^*}(\omega_1 \dots \omega_n H) + \tilde{q}V_{(n+1)\wedge\tau^*}(\omega_1 \dots \omega_n T).
\end{aligned}$$

Again we have the martingale property.

Since the stopped process (4.4.23) is a martingale, we have

$$\begin{aligned}
V_0 &= \widetilde{\mathbb{E}}\left[\frac{1}{(1+r)^{N\wedge\tau^*}} V_{N\wedge\tau^*}\right]\\
&= \widetilde{\mathbb{E}}\left[\mathbb{I}_{\{\tau^*\leq N\}}\frac{1}{(1+r)^{\tau^*}} G_{\tau^*}\right] + \widetilde{\mathbb{E}}\left[\mathbb{I}_{\{\tau^*=\infty\}}\frac{1}{(1+r)^{N}} V_N\right]. \qquad (4.4.24)
\end{aligned}$$

But on those paths for which $\tau^* = \infty$, we must have $V_n > G_n$ for all n and, in particular, $V_N > G_N$. In light of (4.4.10), this can only happen if $G_N < 0$ and $V_N = 0$. Therefore, $\mathbb{I}_{\{\tau^*=\infty\}} V_N = 0$ and (4.4.24) can be simplified to

$$V_0 = \widetilde{\mathbb{E}}\left[\mathbb{I}_{\{\tau^*\leq N\}}\frac{1}{(1+r)^{\tau^*}} G_{\tau^*}\right]. \tag{4.4.25}$$

This is (4.4.22). □

4.5 American Call Options

We saw in Example 4.2.1 that it is sometimes optimal to exercise an American put "early" (i.e., before expiration). For an American call on a non-dividend-paying stock, there is no advantage to early exercise. This is a consequence of Jensen's inequality for conditional expectations, as we show below.

Let $g : [0,\infty) \to \mathbb{R}$ be a convex function satisfying $g(0) = 0$ (see Figure 4.5.1). This means that whenever $s_1 \geq 0$, $s_2 \geq 0$, and $0 \leq \lambda \leq 1$, we have

$$g\big(\lambda s_1 + (1-\lambda)s_2\big) \leq \lambda g(s_1) + (1-\lambda)g(s_2). \tag{4.5.1}$$

For instance, we might have $g(s) = (s-K)^+$, the payoff of a call with strike K.

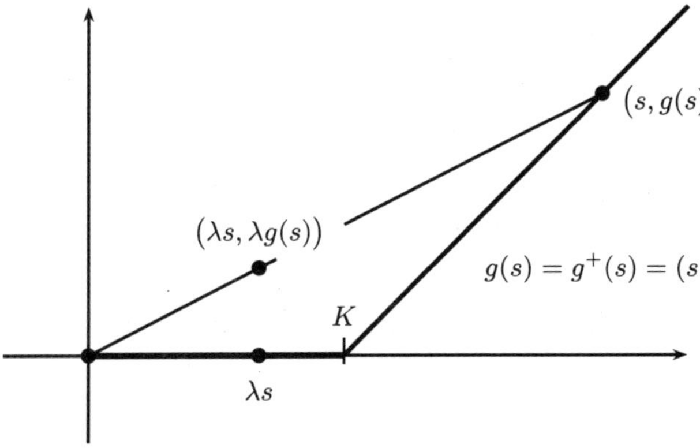

Fig. 4.5.1. Convex function with $g(0) = 0$.

Theorem 4.5.1. *Consider an N-period binomial asset-pricing model with $0 < d < 1 + r < u$ and $r \geq 0$. In this model, consider an American derivative security with convex payoff function $g(s)$ satisfying $g(0) = 0$. The value of this derivative security at time zero, which is (see Definition 4.4.1)*

$$V_0^A = \max_{\tau \in \mathcal{S}_0} \widetilde{\mathbb{E}}\left[\mathbb{I}_{\{\tau \leq N\}} \frac{1}{(1+r)^\tau} g(S_\tau)\right], \tag{4.5.2}$$

is the same as the value of the European derivative security with payoff $g(S_N)$ at expiration N, which is (see Theorem 2.4.7 of Chapter 2)

$$V_0^E = \widetilde{\mathbb{E}}\left[\frac{1}{(1+r)^N} \max\{g(S_N), 0\}\right]. \tag{4.5.3}$$

PROOF: Under our assumptions, g can take negative values. We therefore introduce the function

$$g^+(s) = \max\{g(s), 0\},$$

which takes only nonnegative values and satisfies $g^+(0) = 0$. Furthermore, g^+ is convex. To see this, observe that because g satisfies (4.5.1), we have

$$g(\lambda s_1 + (1-\lambda)s_2) \le \lambda g^+(s_1) + (1-\lambda)g^+(s_2) \text{ for all } s_1 \ge 0,\ s_2 \ge 0,\ \lambda \in [0,1].$$

We also have $0 \le \lambda g^+(s_1) + (1-\lambda)g^+(s_2)$, and hence

$$g^+\big(\lambda s_1 + (1-\lambda)s_2\big) = \max\big\{0, g(\lambda s_1 + (1-\lambda)s_2)\big\} \le \lambda g^+(s_1) + (1-\lambda)g^+(s_2).$$

This last inequality establishes the convexity of g^+. Taking $s_1 = s$ and $s_2 = 0$ in this inequality, we see that

$$g^+(\lambda s) \le \lambda g^+(s) \text{ for all } s \ge 0,\ \lambda \in [0,1]. \tag{4.5.4}$$

Because $\frac{1}{(1+r)^n} S_n$ is a martingale under the risk-neutral probabilities, we have $S_n = \widetilde{\mathbb{E}}_n\left[\frac{1}{1+r} S_{n+1}\right]$ and

$$g^+(S_n) = g^+\left(\widetilde{\mathbb{E}}_n\left[\frac{1}{1+r} S_{n+1}\right]\right).$$

The conditional Jensen's inequality of Theorem 2.3.2(v) of Chapter 2 implies that

$$g^+\left(\widetilde{\mathbb{E}}_n\left[\frac{1}{1+r} S_{n+1}\right]\right) \le \widetilde{\mathbb{E}}_n\left[g^+\left(\frac{1}{1+r} S_{n+1}\right)\right]. \tag{4.5.5}$$

Taking $\lambda = \frac{1}{1+r}$ in (4.5.4), we obtain

$$\widetilde{\mathbb{E}}_n\left[g^+\left(\frac{1}{1+r} S_{n+1}\right)\right] \le \widetilde{\mathbb{E}}_n\left[\frac{1}{1+r} g^+(S_{n+1})\right]. \tag{4.5.6}$$

Putting all of this together, we see that

$$g^+(S_n) \le \widetilde{\mathbb{E}}_n\left[\frac{1}{1+r} g^+(S_{n+1})\right],$$

and multiplication of both sides by $\frac{1}{(1+r)^n}$ yields the submartingale property

$$\frac{1}{(1+r)^n} g^+(S_n) \le \widetilde{\mathbb{E}}_n\left[\frac{1}{(1+r)^{n+1}} g^+(S_{n+1})\right]$$

for the discounted intrinsic value process $\frac{1}{(1+r)^n} g^+(S_n)$. Because this process is a submartingale, Theorem 4.3.3 implies that for every stopping time $\tau \in \mathcal{S}_0$

$$\widetilde{\mathbb{E}}\left[\frac{1}{(1+r)^{N\wedge\tau}} g^+(S_{N\wedge\tau})\right] \le \widetilde{\mathbb{E}}\left[\frac{1}{(1+r)^N} g^+(S_N)\right] = V_0^E. \tag{4.5.7}$$

If $\tau \le N$, then

$$\mathbb{I}_{\{\tau \le N\}} \frac{1}{(1+r)^{\tau}} g(S_\tau) = \frac{1}{(1+r)^{N\wedge\tau}} g(S_{N\wedge\tau}) \le \frac{1}{(1+r)^{N\wedge\tau}} g^+(S_{N\wedge\tau}),$$

and if $\tau = \infty$, then

$$\mathbb{I}_{\{\tau \le N\}} \frac{1}{(1+r)^{\tau}} g(S_\tau) = 0 \le \frac{1}{(1+r)^{N\wedge\tau}} g^+(S_{N\wedge\tau}).$$

In either case, we have the same result and so

$$\widetilde{\mathbb{E}}\left[\mathbb{I}_{\{\tau \le N\}} \frac{1}{(1+r)^{\tau}} g(S_\tau)\right] \le \widetilde{\mathbb{E}}\left[\frac{1}{(1+r)^{N\wedge\tau}} g^+(S_{N\wedge\tau})\right] \le V_0^E,$$

where we have used (4.5.7) for the last step. Since this last inequality holds for every $\tau \in \mathcal{S}_0$, we must have

$$V_0^A = \max_{\tau \in \mathcal{S}_0} \widetilde{\mathbb{E}}\left[\mathbb{I}_{\{\tau \le N\}} \frac{1}{(1+r)^{\tau}} g(S_\tau)\right] \le V_0^E. \qquad \square$$

Theorem 4.5.1 shows that the early exercise feature of the American call contributes nothing to its value. An examination of the proof of the theorem indicates that this is because the discounted intrinsic value of the call is a submartingale (i.e., has a tendency to rise) under the risk-neutral probabilities. The discounted intrinsic value of an American put is not a submartingale. If $g^+(s) = (K-s)^+$, then the Jensen inequality (4.5.5) still holds but (4.5.6) does not. Jensen's inequality says that the convex payoff of the put imparts to the discounted intrinsic value of the put a tendency to rise over time, but this may be overcome by a second effect. Because the owner of the put receives K upon exercise, she may exercise early in order to prevent the value of this payment from being discounted away. For low stock prices, this second effect becomes more important than the convexity, and early exercise becomes optimal. For the call, the owner pays K and prefers that the value of this payment be discounted away before exercise. This reinforces the convexity effect and makes early exercise undesirable.

4.6 Summary

Unlike a European derivative security, which can only be exercised at one time, the so-called *expiration date*, an American derivative security entitles its owner to exercise at any time prior to or at the expiration date. One consequence of this is that the value of an American derivative security is always at least as great as the payoff its owner would receive from immediate exercise, the so-called *intrinsic value*. Unlike a European derivative security, whose discounted value is a martingale under the risk-neutral measure, the discounted value of an American derivative security is a supermartingale under the risk-neutral measure. It has a tendency to go down at exactly those moments when it should be exercised. In fact, the value process of an American derivative security is the smallest nonnegative process dominating the intrinsic value and that is a supermartingale under the risk-neutral measure, when discounted. This is the content of Theorem 4.4.2. That theorem leads to the American derivative security pricing algorithm of Theorem 4.4.3:

$$V_N(\omega_1 \dots \omega_N) = \max\{G_N(\omega_1 \dots \omega_N), 0\}, \tag{4.4.10}$$

$$V_n(\omega_1 \dots \omega_n) = \max\Bigg\{G_n(\omega_1 \dots \omega_n), \tag{4.4.11}$$

$$\frac{1}{1+r}\Big[\tilde{p}V_{n+1}(\omega_1 \dots \omega_n H) + \tilde{q}V_{n+1}(\omega_1 \dots \omega_n T)\Big]\Bigg\},$$

for $n = N-1, \dots, 0$. Here $G_n(\omega_1 \dots \omega_n)$ is the intrinsic value of the security at time n if the first n coin tosses result in $\omega_1 \dots \omega_n$.

The supermartingale property for the discounted American derivative security value process permits an agent holding a short position in the security to construct a hedge. This hedge is constructed by the same formula used for a European derivative security:

$$\Delta_n(\omega_1 \dots \omega_n) = \frac{V_{n+1}(\omega_1 \dots \omega_n H) - V_{n+1}(\omega_1 \dots \omega_n T)}{S_{n+1}(\omega_1 \dots \omega_n H) - S_{n+1}(\omega_1 \dots \omega_n T)}. \tag{4.4.14}$$

In those periods n in which the discounted value process has a strictly downward trend, the short position hedger can consume

$$C_n(\omega_1 \dots \omega_n) = V_n(\omega_1 \dots \omega_n) - \frac{1}{1+r}\Big[\tilde{p}V_{n+1}(\omega_1 \dots \omega_n H) + \tilde{q}V_{n+1}(\omega_1 \dots \omega_n T)\Big] \tag{4.4.15}$$

and still maintain the hedge. This is the content of Theorem 4.4.4.

The owner of an American derivative security should exercise the first time the value of the security agrees with its intrinsic value. This exercise rule results in a payment whose discounted risk-neutral value at time zero agrees with the price of the American derivative security at time zero (i.e., this rule permits the owner to capture the full value of the American derivative security). This is the content of Theorem 4.4.5.

The proofs of the claims above use the idea of a stopping time, a random time that makes the decision to stop (exercise) without looking ahead; see Definition 4.3.1. The exercise strategy of the owner of an American derivative security should be a stopping time (i.e., it may depend on past stock price movements but must make the decision to exercise without looking at future price movements). Once a stopping time is chosen, one can compute the risk-neutral expected discounted payoff of the derivative security when that stopping time is used. The value of an American derivative security is the maximum of these risk-neutral expected discounted payoffs over all stopping times. This is Definition 4.4.1.

Martingales, supermartingale, and submartingales evaluated at stopping times rather than nonrandom times have the same trends as if they were evaluated at random times. In particular, if X_n, $n = 0, 1, \ldots, N$ is a submartingale under $\widetilde{\mathbb{P}}$ and τ is a stopping time, then $X_{n \wedge \tau}$, $n = 0, 1, \ldots, N$ is also a submartingale under $\widetilde{\mathbb{P}}$ and $\widetilde{\mathbb{E}} X_{n \wedge \tau} \leq \widetilde{\mathbb{E}} X_n$. Results of this type are called *optional sampling*. Using optional sampling, one can show that the value of an American call on a stock paying no dividends is the same as the value of a European call on the same stock (i.e., the early exercise option in the American call has zero value). This is the content of Theorem 4.5.1 .

4.7 Notes

A rigorous analysis of American derivative securities based on stopping times was initiated by Bensoussan [2] and continued by Karatzas [26]. A comprehensive treatment appears in Karatzas and Shreve [28]. All of this material treats continuous-time models; it has been specialized to the binomial model in this text.

4.8 Exercises

Exercise 4.1. In the three-period model of Figure 1.2.2 of Chapter 1, let the interest rate be $r = \frac{1}{4}$ so the risk-neutral probabilities are $\tilde{p} = \tilde{q} = \frac{1}{2}$.

(i) Determine the price at time zero, denoted V_0^P, of the American put that expires at time three and has intrinsic value $g_P(s) = (4 - s)^+$.
(ii) Determine the price at time zero, denoted V_0^C, of the American call that expires at time three and has intrinsic value $g_C(s) = (s - 4)^+$.
(iii) Determine the price at time zero, denoted V_0^S, of the American straddle that expires at time three and has intrinsic value $g_S(s) = g_P(s) + g_C(s)$.
(iv) Explain why $V_0^S < V_0^P + V_0^C$.

Exercise 4.2. In Example 4.2.1, we computed the time-zero value of the American put with strike price 5 to be 1.36. Consider an agent who borrows 1.36 at time zero and buys the put. Explain how this agent can generate sufficient funds to pay off his loan (which grows by 25% each period) by trading in the stock and money markets and optimally exercising the put.

Exercise 4.3. In the three-period model of Figure 1.2.2 of Chapter 1, let the interest rate be $r = \frac{1}{4}$ so the risk-neutral probabilities are $\tilde{p} = \tilde{q} = \frac{1}{2}$. Find the time-zero price and optimal exercise policy (optimal stopping time) for the path-dependent American derivative security whose intrinsic value at each time n, $n = 0, 1, 2, 3$, is $\left(4 - \frac{1}{n+1}\sum_{j=0}^{n} S_j\right)^+$. This intrinsic value is a put on the average stock price between time zero and time n.

Exercise 4.4. Consider the American put of Example 4.2.1, which has strike price 5. Suppose at time zero we sell this put to a purchaser who has inside information about the stock movements and uses the exercise rule ρ of (4.3.2). In particular, if the first toss is going to result in H, the owner of the put exercises at time zero, when the put has intrinsic value 1. If the first toss results in T and the second toss is going to result in H, the owner exercises at time one, when the put has intrinsic value 3. If the first two tosses result in TT, the owner exercises at time two, when the intrinsic value is 4. In summary, the owner of the put has the payoff random variable

$$Y(HH) = 1,\ Y(HT) = 1,\ Y(TH) = 3,\ Y(TT) = 4. \tag{4.8.1}$$

The risk-neutral expected value of this payoff, discounted from the time of payment back to zero, is

$$\widetilde{\mathbb{E}}\left[\left(\frac{4}{5}\right)^{\rho} Y\right] = \frac{1}{4}\left[1 + 1 + \frac{4}{5}\cdot 3 + \frac{16}{25}\cdot 4\right] = 1.74. \tag{4.8.2}$$

The time-zero price of the put computed in Example 4.2.1 is only 1.36. Do we need to charge the insider more than this amount if we are going to successfully hedge our short position after selling the put to her? Explain why or why not.

Exercise 4.5. In equation (4.4.5), the maximum is computed over all stopping times in $\mathcal{S}_0$. List all the stopping times in $\mathcal{S}_0$ (there are 26), and from among them, list the stopping times that never exercise when the option is out of the money (there are 11). For each stopping time τ in the latter set, compute $\mathbb{E}\left[\mathbb{I}_{\{\tau\le 2\}}\left(\frac{4}{5}\right)^{\tau} G_\tau\right]$. Verify that the largest value for this quantity is given by the stopping time of (4.4.6), the one that makes this quantity equal to the 1.36 computed in (4.4.7).

Exercise 4.6 (Estimating American put prices). For each n, where $n = 0, 1, \ldots, N$, let G_n be a random variable depending on the first n coin tosses. The time-zero value of a derivative security that can be exercised at any time $n \le N$ for payoff G_n but *must be exercised at time N if it has not been exercised before that time* is

$$V_0 = \max_{\tau\in\mathcal{S}_0, \tau\le N} \widetilde{\mathbb{E}}\left[\frac{1}{(1+r)^{\tau}} G_\tau\right]. \tag{4.8.3}$$

In contrast to equation (4.4.1) in the Definition 4.4.1 for American derivative securities, here we consider only stopping times that take one of the values $0, 1, \ldots, N$ and not the value ∞.

(i) Consider $G_n = K - S_n$, the derivative security that permits its owner to sell one share of stock for payment K at any time up to and including N, but if the owner does not sell by time N, then she must do so at time N. Show that the optimal exercise policy is to sell the stock at time zero and that the value of this derivative security is $K - S_0$. (Assume $r \geq 0$.)

(ii) Explain why a portfolio that holds the derivative security in (i) and a European call with strike K and expiration time N is at least as valuable as an American put struck at K with expiration time N. Denote the time-zero value of the European call by V_0^{EC} and the time-zero value of the American put by V_0^{AP}. Conclude that the upper bound

$$V_0^{AP} \leq K - S_0 + V_0^{EC} \tag{4.8.4}$$

on V_0^{AP} holds.

(iii) Use put–call parity (Exercise 2.11 of Chapter 2) to derive the lower bound on V_0^{AP}:

$$\frac{K}{(1+r)^N} - S_0 + V_0^{EC} \leq V_0^{AP}. \tag{4.8.5}$$

Exercise 4.7. For the class of derivative securities described in Exercise 4.6 whose time-zero price is given by (4.8.3), let $G_n = S_n - K$. This derivative security permits its owner to buy one share of stock in exchange for a payment of K at any time up to the expiration time N. If the purchase has not been made at time N, it must be made then. Determine the time-zero value and optimal exercise policy for this derivative security. (Assume $r \geq 0$.)

5

Random Walk

5.1 Introduction

In this section, we consider a symmetric random walk, which is the discrete-time version of Brownian motion, introduced in Chapter 3 of Volume II. We derive several properties of a random walk, and shall ultimately see that Brownian motion has similar properties. In particular, in this chapter we consider *first passage times* and the *reflection principle* for a symmetric random walk. For Brownian motion, these concepts are used in the computation of the price of a variety of exotic options.

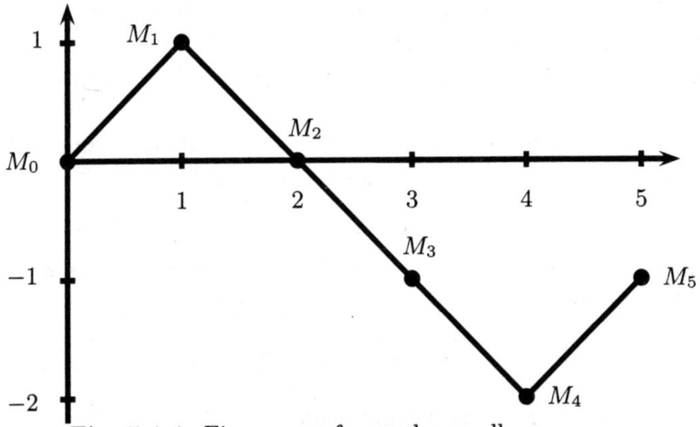

Fig. 5.1.1. Five steps of a random walk

To construct a symmetric random walk, we repeatedly toss a fair coin (p, the probability of H on each toss, and $q = 1 - p$, the probability of T on each toss, are both equal to $\frac{1}{2}$). We denote the successive outcomes of the tosses by $\omega_1\omega_2\omega_3\ldots$. Let

$$X_j = \begin{cases} 1, & \text{if } \omega_j = H, \\ -1, & \text{if } \omega_j = T, \end{cases} \tag{5.1.1}$$

and define $M_0 = 0$,

$$M_n = \sum_{j=1}^{n} X_j, \ n = 1, 2, \dots . \tag{5.1.2}$$

The process M_n, $n = 0, 1, 2, \dots$ is a *symmetric random walk.* With each toss, it either steps up one unit or down one unit, and each of the two possibilities is equally likely. If $p \neq \frac{1}{2}$, we would still have a random walk, but it would be *asymmetric.* In other words, the symmetric and asymmetric random walks have the same set of possible paths; they differ only in the assignment of probabilities to these paths.

The symmetric random walk is both a martingale and a Markov process.

5.2 First Passage Times

The symmetric random walk of Section 5.1 starts at 0 at time zero. Fix an integer m, and let τ_m denote the first time the random walk reaches level m; i.e.,

$$\tau_m = \min\{n; M_n = m\}. \tag{5.2.1}$$

If the random walk never reaches the level m, we define τ_m to be infinity.

The random variable τ_m is a stopping time, called the *first passage time* of the random walk to level m. We shall determine its distribution. We shall see that τ_m is finite with probability 1 (i.e., with probability 1, the random walk eventually reaches the level m), but $\mathbb{E}\tau_m = \infty$. Once we have determined the distribution of τ_m for a symmetric random walk, we shall see how to modify the formulas to obtain information about the distribution of τ_m when the walk is asymmetric.

Our study of the distribution of τ_m for a symmetric random walk uses the martingale (5.2.2) below, which is discussed in Exercise 2.4(ii) at the end of Chapter 2.

Lemma 5.2.1. *Let M_n be a symmetric random walk. Fix a number σ and define the process*

$$S_n = e^{\sigma M_n} \left(\frac{2}{e^{\sigma} + e^{-\sigma}} \right)^n . \tag{5.2.2}$$

Then S_n, $n = 0, 1, 2, \dots$ is a martingale.

PROOF: In the notation of (5.1.1) and (5.1.2), we have

$$S_{n+1} = S_n \left(\frac{2}{e^{\sigma} + e^{-\sigma}} \right) e^{\sigma X_{n+1}} .$$

We take out what is known (Theorem 2.3.2(ii)) and use independence (Theorem 2.3.2(iv)) to write

$$\begin{aligned}\mathbb{E}_n S_{n+1} &= S_n \left(\frac{2}{e^{\sigma}+e^{-\sigma}}\right) \mathbb{E} e^{\sigma X_{n+1}} \\ &= S_n \left(\frac{2}{e^{\sigma}+e^{-\sigma}}\right)\left(\frac{1}{2}e^{\sigma}+\frac{1}{2}e^{-\sigma}\right) \\ &= S_n.\end{aligned}$$

This shows that the process S_n, $n = 0, 1, 2, \ldots$ is a martingale. □

Because a martingale stopped at a stopping time is still a martingale (Theorem 4.3.2), the process $S_{n\wedge\tau_m}$ is a martingale and hence has constant expectation, i.e.,

$$1 = S_0 = \mathbb{E} S_{n\wedge\tau_m} = \mathbb{E}\left[e^{\sigma M_{n\wedge\tau_m}}\left(\frac{2}{e^{\sigma}+e^{-\sigma}}\right)^{n\wedge\tau_m}\right] \text{ for all } n \geq 0. \quad (5.2.3)$$

We would like to let $n \to \infty$ in (5.2.3). In order to do that, we must determine the limit as $n \to \infty$ of $e^{\sigma M_{n\wedge\tau_m}}\left(\frac{2}{e^{\sigma}+e^{-\sigma}}\right)^{n\wedge\tau_m}$. We treat the two factors separately.

We observe first that $\cosh(\sigma) = \frac{e^{\sigma}+e^{-\sigma}}{2}$ attains its minimum at $\sigma = 0$, where it takes the value 1. Hence, $\cosh(\sigma) > 1$ for all $\sigma > 0$, so

$$0 < \frac{2}{e^{\sigma}+e^{-\sigma}} < 1 \text{ for all } \sigma > 0. \quad (5.2.4)$$

We now fix $\sigma > 0$ and conclude from (5.2.4) that

$$\lim_{n\to\infty}\left(\frac{2}{e^{\sigma}+e^{-\sigma}}\right)^{n\wedge\tau_m} = \begin{cases}\left(\frac{2}{e^{\sigma}+e^{-\sigma}}\right)^{\tau_m}, & \text{if } \tau_m < \infty, \\ 0, & \text{if } \tau_m = \infty.\end{cases} \quad (5.2.5)$$

We may write the right-hand side of (5.2.5) as $\mathbb{I}_{\{\tau_m<\infty\}}\left(\frac{2}{e^{\sigma}+e^{-\sigma}}\right)^{\tau_m}$, which captures both cases.

To study the other factor, $e^{\sigma M_{n\wedge\tau_m}}$, we assume that $m > 0$ and note that $M_{n\wedge\tau_m} \leq m$ because we stop this martingale when it reaches the level m. Hence, regardless of whether τ_m is finite or infinite, we have

$$0 \leq e^{\sigma M_{n\wedge\tau_m}} \leq e^{\sigma m}. \quad (5.2.6)$$

We also have

$$\lim_{n\to\infty} e^{\sigma M_{n\wedge\tau_m}} = e^{\sigma M_{\tau_m}} = e^{\sigma m} \text{ if } \tau_m < \infty. \quad (5.2.7)$$

Taking the product of (5.2.5) and (5.2.7), we see that

$$\lim_{n\to\infty} e^{\sigma M_{n\wedge\tau_m}}\left(\frac{2}{e^{\sigma}+e^{-\sigma}}\right)^{n\wedge\tau_m} = e^{\sigma m}\left(\frac{2}{e^{\sigma}+e^{-\sigma}}\right)^{\tau_m} \text{ if } \tau_m < \infty. \quad (5.2.8)$$

We don't know that $e^{\sigma M_{n\wedge\tau_m}}$ has a limit as $n \to \infty$ along the paths for which $\tau_m = \infty$, but that does not matter because this term is bounded (see (5.2.6)) and $\left(\frac{2}{e^\sigma+e^{-\sigma}}\right)^{n\wedge\tau_m}$ has limit zero. Hence,

$$\lim_{n\to\infty} e^{\sigma M_{n\wedge\tau_m}} \left(\frac{2}{e^\sigma + e^{-\sigma}}\right)^{n\wedge\tau_m} = 0 \text{ if } \tau_m = \infty. \tag{5.2.9}$$

It follows from (5.2.8) and (5.2.9) that

$$\lim_{n\to\infty} e^{\sigma M_{n\wedge\tau_m}} \left(\frac{2}{e^\sigma + e^{-\sigma}}\right)^{n\wedge\tau_m} = \mathbb{I}_{\{\tau_m<\infty\}} e^{\sigma m} \left(\frac{2}{e^\sigma + e^{-\sigma}}\right)^{\tau_m}. \tag{5.2.10}$$

We may now take the limit in (5.2.3) to obtain[1]

$$\mathbb{E}\left[\mathbb{I}_{\{\tau_m<\infty\}} e^{\sigma m} \left(\frac{2}{e^\sigma + e^{-\sigma}}\right)^{\tau_m}\right] = 1. \tag{5.2.11}$$

Equation (5.2.11) was derived under the assumption that σ is strictly positive, the assumption we used to derive (5.2.5). Equation (5.2.11) thus holds for all strictly positive σ. We may not set $\sigma = 0$ in (5.2.11), but we may compute the limit of both sides[2] as $\sigma \downarrow 0$, and this yields $\mathbb{E}\mathbb{I}_{\{\tau_m<\infty\}} = 1$, i.e.,

$$\mathbb{P}\{\tau_m < \infty\} = 1. \tag{5.2.12}$$

This tells us that, with probability 1, the symmetric random walk reaches the level m. Nonetheless, there are paths of the symmetric random walk that never reach level m. For example, any path in which the tails obtained at any time always outnumber the heads (such a path might begin with $TTHTTHTHHT$...) will begin at zero and take only strictly negative values thereafter. Such a path never reaches the positive level m. Equation (5.2.12) asserts that although there are infinitely many such paths (in fact, uncountably infinitely many), taken all together, the set of these paths has zero probability. We have assumed for this discussion that m is strictly positive (we used this assumption to derive (5.2.6)), but because of the symmetry of the random walk, the conclusion (5.2.12) also holds if m is strictly negative.

When some event happens with probability 1, we say the event happens *almost surely*. We have proved the following theorem.

Theorem 5.2.2. *Let m be an arbitrary nonzero integer. The symmetric random walk reaches the level m almost surely; i.e., the first passage time τ_m to level m is finite almost surely.*

---

[1] It is not always possible to conclude from convergence of a sequence of random variables that the expectations converge in the same way. In this particular case, we may take this step because the random variables appearing in (5.2.10) are bounded between two constants, 0 and $e^{\sigma m}$. The applicable theorem is the Dominated Convergence Theorem 1.4.9 of Volume II, Chapter 1.

[2] The computation of the limit of the left-hand side of (5.2.11) requires another application of the Dominated Convergence Theorem 1.4.9 of Volume II, Chapter 1.

In order to determine the distribution of the first passage time τ_m, we examine its moment-generating function $\varphi_{\tau_m}(u) = \mathbb{E}e^{u\tau_m}$. For all $x \geq 0$, $e^x = 1 + x + \frac{1}{2}x^2 + \cdots \geq x$. Hence, for positive u, we have $e^{u\tau_m} \geq u\tau_m$ and so $\varphi_{\tau_m}(u) = \mathbb{E}e^{u\tau_m} \geq u\mathbb{E}\tau_m$. We shall see in Corollary 5.2.4 below that $\mathbb{E}\tau_m = \infty$, which implies that $\varphi_{\tau_m}(u) = \infty$ for $u > 0$. For $u = 0$, we have $\varphi_{\tau_m}(u) = 1$. Therefore, the moment-generating function $\varphi_{\tau_m}(u)$ is interesting only for $u < 0$. For $u < 0$, we set $\alpha = e^u$ so that $0 < \alpha < 1$ and $\varphi_{\tau_m}(u) = \mathbb{E}\alpha^{\tau_m}$.

Theorem 5.2.3. *Let m be a nonzero integer. The first passage time τ_m for the symmetric random walk satisfies*

$$\mathbb{E}\alpha^{\tau_m} = \left(\frac{1 - \sqrt{1-\alpha^2}}{\alpha}\right)^{|m|} \quad \textit{for all } \alpha \in (0,1). \tag{5.2.13}$$

PROOF: For the symmetric random walk, τ_m and τ_{-m} have the same distribution, so it is enough to prove the theorem for the case that m is a positive integer. We take m to be a positive integer. Because $\mathbb{P}\{\tau_m < \infty\} = 1$, we may simplify (5.2.11) to

$$\mathbb{E}\left[e^{\sigma m}\left(\frac{2}{e^{\sigma} + e^{-\sigma}}\right)^{\tau_m}\right] = 1. \tag{5.2.14}$$

This equation holds for all strictly positive σ.

To obtain (5.2.13) from (5.2.14), we let $\alpha \in (0,1)$ be given and solve for $\sigma > 0$, which satisfies

$$\alpha = \frac{2}{e^{\sigma} + e^{-\sigma}}. \tag{5.2.15}$$

This is equivalent to

$$\alpha e^{\sigma} + \alpha e^{-\sigma} - 2 = 0,$$

which, when divided by e^{σ}, leads to

$$\alpha\left(e^{-\sigma}\right)^2 - 2e^{-\sigma} + \alpha = 0.$$

This last equation is a quadratic equation in the unknown $e^{-\sigma}$, and the solutions are

$$e^{-\sigma} = \frac{2 \pm \sqrt{4 - 4\alpha^2}}{2\alpha} = \frac{1 \pm \sqrt{1-\alpha^2}}{\alpha}.$$

We need to find a strictly positive σ satisfying this equation, and thus we need $e^{-\sigma}$ to be strictly less than one. That suggests we should take the solution for $e^{-\sigma}$ corresponding to the negative sign in the formula above; i.e.,

$$e^{-\sigma} = \frac{1 - \sqrt{1-\alpha^2}}{\alpha}. \tag{5.2.16}$$

We verify that (5.2.16) leads to a value of σ that is strictly positive. Because we have chosen α to satisfy $0 < \alpha < 1$, we have

$$0 < (1-\alpha)^2 < 1-\alpha < 1-\alpha^2.$$

Taking positive square roots, we obtain

$$1-\alpha < \sqrt{1-\alpha^2}.$$

Therefore, $1-\sqrt{1-\alpha^2} < \alpha$, and dividing through by α, we see that the right-hand side of (5.2.16) is strictly less than 1. This implies that σ in (5.2.16) is strictly positive.

With σ and α related by (5.2.16), we have that α and σ are also related by (5.2.15) and hence may rewrite (5.2.14) as

$$\mathbb{E}\left[\left(\frac{\alpha}{1-\sqrt{1-\alpha^2}}\right)^m \alpha^{\tau_m}\right] = 1.$$

Because $\left(\frac{\alpha}{1-\sqrt{1-\alpha^2}}\right)^m$ is not random, we may take it outside the expectation and divide through by it. Equation (5.2.13) for positive m follows immediately.
□

Corollary 5.2.4. *Under the conditions of Theorem 5.2.3, we have*

$$\mathbb{E}\tau_m = \infty. \qquad (5.2.17)$$

PROOF:[3] We first show that $\mathbb{E}\tau_1 = \infty$. To do this, we differentiate both sides of (5.2.13) with respect to α:

$$\begin{aligned}
\mathbb{E}\left[\tau_1\alpha^{\tau_1-1}\right] &= \frac{\partial}{\partial\alpha}\mathbb{E}\alpha^{\tau_1}\\
&= \frac{\partial}{\partial\alpha}\frac{1-\sqrt{1-\alpha^2}}{\alpha}\\
&= \frac{-\frac{1}{2}(1-\alpha^2)^{-\frac{1}{2}}(-2\alpha)\alpha - \left(1-(1-\alpha^2)^{\frac{1}{2}}\right)}{\alpha^2}\\
&= \frac{\alpha^2(1-\alpha^2)^{-\frac{1}{2}} - 1 + (1-\alpha^2)^{\frac{1}{2}}}{\alpha^2}\\
&= \frac{\alpha^2 - \sqrt{1-\alpha^2} + 1 - \alpha^2}{\alpha^2\sqrt{1-\alpha^2}}\\
&= \frac{1-\sqrt{1-\alpha^2}}{\alpha^2\sqrt{1-\alpha^2}}.
\end{aligned}$$

This equation is valid for all $\alpha \in (0,1)$. We may not substitute $\alpha = 1$ into this equation, but we may take the limit of both sides as $\alpha \uparrow 1$, and this gives us $\mathbb{E}\tau_1 = \infty$.

[3] There are two steps in the proof of Corollary 5.2.4 that require the interchange of limit and expectation. The first of these, differentiation of $\mathbb{E}\alpha^{\tau_1}$ with respect to α, can be justified by an argument like that in Exercise 8 in Volume II, Chapter 1. The second, where we let $\alpha \uparrow 1$, is an application of the Monotone Convergence Theorem 1.4.5 of Volume II, Chapter 1.

For $m \geq 1$, we have $\tau_m \geq \tau_1$ and hence $\mathbb{E}\tau_m \geq \mathbb{E}\tau_1 = \infty$. For strictly negative integers m, the symmetry of the random walk now implies $\mathbb{E}\tau_m = \infty$.
□

From the formula (5.2.13), it is possible to compute explicitly the distribution of the random variable τ_1. We have the special case of formula (5.2.13):

$$\mathbb{E}\alpha^{\tau_1} = \frac{1-\sqrt{1-\alpha^2}}{\alpha} \text{ for all } \alpha \in (0,1). \tag{5.2.18}$$

Because the random walk can reach level 1 only on an odd-numbered step, the left-hand side of (5.2.18) may be rewritten as

$$\mathbb{E}\alpha^{\tau_1} = \sum_{j=1}^{\infty} \alpha^{2j-1}\mathbb{P}\{\tau_1 = 2j-1\} \tag{5.2.19}$$

We work out a power series expansion for the right-hand side of (5.2.18). Define $f(x) = 1 - \sqrt{1-x}$ so that

$$f'(x) = \frac{1}{2}(1-x)^{-\frac{1}{2}},$$
$$f''(x) = \frac{1}{4}(1-x)^{-\frac{3}{2}},$$
$$f'''(x) = \frac{3}{8}(1-x)^{-\frac{5}{2}},$$

and, in general, the jth-order derivative of f is

$$f^{(j)}(x) = \frac{1\cdot 3 \cdots (2j-3)}{2^j}(1-x)^{-\frac{2j-1}{2}}, \quad j = 1,2,3,\ldots.$$

Evaluating at 0, we obtain

$$f(0) = 0, \; f'(0) = \frac{1}{2}, \; f''(0) = \frac{1}{4}, \; f'''(0) = \frac{3}{8}$$

and, in general,

$$\begin{aligned} f^{(j)}(0) &= \frac{1\cdot 3 \ldots (2j-3)}{2^j} \\ &= \frac{1\cdot 3 \ldots (2j-3)}{2^j} \cdot \frac{2\cdot 4 \cdots (2j-2)}{2^{j-1}(j-1)!} \\ &= \left(\frac{1}{2}\right)^{2j-1} \frac{(2j-2)!}{(j-1)!}, \quad j = 1,2,3,\ldots. \end{aligned} \tag{5.2.20}$$

(We use here the definition $0! = 1$.) The Taylor series expansion of $f(x)$ is

$$f(x) = 1 - \sqrt{1-x} = \sum_{j=0}^{\infty} \frac{1}{j!} f^{(j)}(0)x^j = \sum_{j=1}^{\infty} \left(\frac{1}{2}\right)^{2j-1} \frac{(2j-2)!}{j!(j-1)!} x^j.$$

Therefore,

$$\frac{1-\sqrt{1-\alpha^2}}{\alpha} = \frac{f(\alpha^2)}{\alpha} = \sum_{j=1}^{\infty}\left(\frac{\alpha}{2}\right)^{2j-1}\frac{(2j-2)!}{j!(j-1)!}. \tag{5.2.21}$$

We have thus worked out the power series (5.2.19) and (5.2.21) for the two sides of (5.2.18). Equating them, we obtain

$$\sum_{j=1}^{\infty}\alpha^{2j-1}\mathbb{P}\{\tau_1 = 2j-1\} = \sum_{j=1}^{\infty}\left(\frac{\alpha}{2}\right)^{2j-1}\frac{(2j-2)!}{j!(j-1)!} \quad \text{for all } \alpha \in (0,1).$$

The only way these two power series can be equal for all $\alpha \in (0,1)$ is for their coefficients to be equal term-by-term (i.e., the term multiplying α^{2j-1} must be the same in both series). This gives us the formula

$$\mathbb{P}\{\tau_1 = 2j-1\} = \frac{(2j-2)!}{j!(j-1)!}\cdot\left(\frac{1}{2}\right)^{2j-1}, \quad j = 1, 2, \ldots. \tag{5.2.22}$$

We verify (5.2.22) for the first few values of j. For $j = 1$, we have

$$\mathbb{P}\{\tau_1 = 1\} = \frac{0!}{1!0!}\cdot\frac{1}{2} = \frac{1}{2}.$$

The only way τ_1 can be 1 is for the first coin toss to result in H, and the probability of this for a symmetric random walk is $\frac{1}{2}$. For $j = 2$, we have

$$\mathbb{P}\{\tau_1 = 3\} = \frac{2!}{2!1!}\cdot\left(\frac{1}{2}\right)^3 = \left(\frac{1}{2}\right)^3.$$

The only way τ_1 can be 3 is for the first three coin tosses to result in THH, and the probability of this is $\left(\frac{1}{2}\right)^3$. For $j = 3$, we have

$$\mathbb{P}\{\tau_1 = 5\} = \frac{4!}{3!2!}\cdot\left(\frac{1}{2}\right)^5 = 2\cdot\left(\frac{1}{2}\right)^5.$$

There are two ways τ_1 can be 5; the first five tosses could be either $THTHH$ or $TTHHH$, and the probability of each of these outcomes is $\left(\frac{1}{2}\right)^5$.

From these examples, we see how to interpret the two factors appearing on the right-hand side of (5.2.22). The term $\frac{(2j-2)!}{j!(j-1)!}$ counts the number of paths with $2j-1$ steps that first reach level 1 on the $(2j-1)$st step so that $\tau_1 = 2j-1$. The term $\left(\frac{1}{2}\right)^{2j-1}$ is the probability of each of these paths. Suppose now that the random walk is not symmetric, but has probability p for H and probability $q = 1-p$ for T. The number of paths that first reach level 1 on the $(2j-1)$st step is unaffected. However, since each of these paths must have j up steps and $j-1$ down steps, the probability of such a path is now $p^j q^{j-1}$. This observation leads to the following theorem.

Theorem 5.2.5. *Let τ_1 be the first passage time to level 1 of a random walk that has probability p for an up step and probability $q = 1-p$ for a down step. Then*

$$\mathbb{P}\{\tau_1 = 2j-1\} = \frac{(2j-2)!}{j!(j-1)!}p^j q^{j-1}, \quad j = 1, 2, \dots. \tag{5.2.23}$$

5.3 Reflection Principle

In this section, we give a second proof of Theorem 5.2.5, based on the *reflection principle.* We shall use this same idea in our study of Brownian motion. As in Section 5.2, we first consider a symmetric random walk, obtaining formula (5.2.22). The remainder of the argument to obtain Theorem 5.2.5 is the same as in Section 5.2.

Suppose we toss a coin an odd number $(2j-1)$ of times. Some of the paths of the random walk will reach the level 1 in the first $2j-1$ steps and others will not. In the case of three tosses, there are eight possible paths and five of these reach the level 1 (see Figure 5.3.1). Consider a path that reaches the level 1 at some time $\tau_1 \le 2j-1$. From that moment on, we can create a "reflected" path, which steps up each time the original path steps down and steps down each time the original path steps up. If the original path ends above 1 at the final time $2j-1$, the reflected path ends below 1, and vice versa. If the original path ends at 1, the reflected path does also.

To count the number of paths that reach the level 1 by time $2j-1$, we can count the number of paths that exceed 1 at time $2j-1$, the number of paths that are at 1 at time $2j-1$, and the number of reflected paths that exceed 1 at time $2j-1$. The reflected paths that exceed 1 correspond to paths that reached 1 at some time prior to $2j-1$ but are below 1 at time $2j-1$. In Figure 5.3.1, there is only one path that exceeds 1 at time three: HHH. There are three paths that are at 1 at time three: HHT, HTH, and THH. There is one path whose reflected path exceeds 1 at time three: HTT. These account for all five paths that reach level 1 by time three.

There are as many reflected paths that exceed 1 at time $2j-1$ as there are original paths that exceed 1 at time $2j-1$. In Figure 5.3.1, there is one of each of these. Thus, to count the number of paths that reach level 1 by time $2j-1$, we can count the paths that are at 1 at time $2j-1$ and then add on *twice* the number of paths that exceed 1 at time $2j-1$. In other words, for the symmetric random walk,

$$\mathbb{P}\{\tau_1 \le 2j-1\} = \mathbb{P}\{M_{2j-1} = 1\} + 2\mathbb{P}\{M_{2j-1} \ge 3\}.$$

But for the symmetric random walk, $\mathbb{P}\{M_{2j-1} \ge 3\} = \mathbb{P}\{M_{2j-1} \le -3\}$, so

$$\begin{aligned}\mathbb{P}\{\tau_1 \le 2j-1\} &= \mathbb{P}\{M_{2j-1} = 1\} + \mathbb{P}\{M_{2j-1} \ge 3\} + \mathbb{P}\{M_{2j-1} \le -3\}\\ &= 1 - \mathbb{P}\{M_{2j-1} = -1\}.\end{aligned}$$

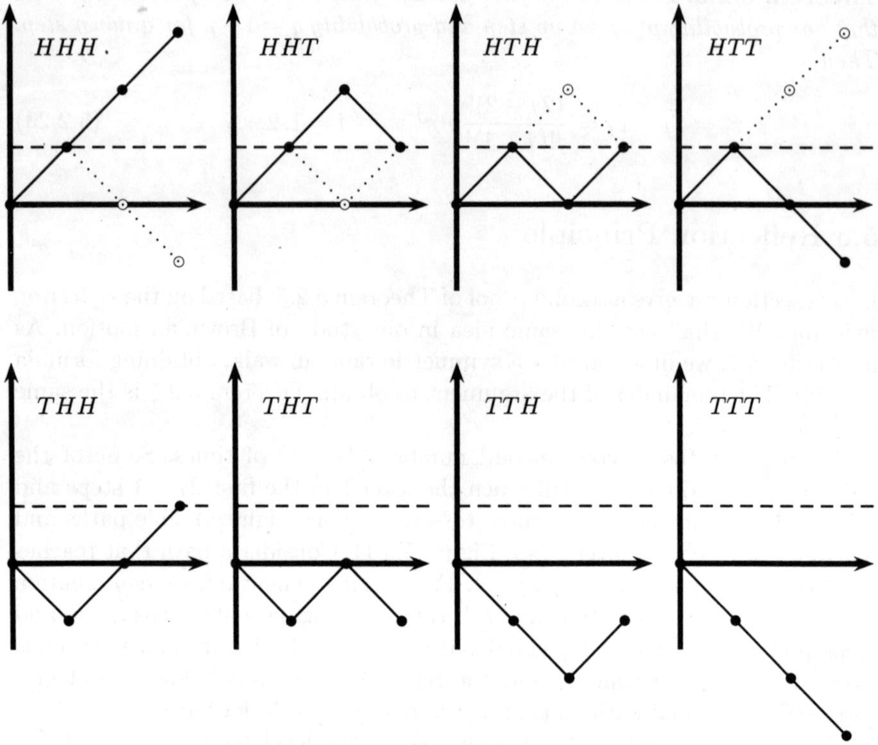

Fig. 5.3.1. Eight paths and four reflected paths.

In order for M_{2j-1} to be -1, in the first $2j-1$ tosses there must be $j-1$ heads and j tails. There are

$$\binom{2j-1}{j} = \frac{(2j-1)!}{j!(j-1)!}$$

paths that have this property, and each such path has probability $\left(\frac{1}{2}\right)^{2j-1}$. Hence,

$$\mathbb{P}\{M_{2j-1} = -1\} = \left(\frac{1}{2}\right)^{2j-1} \frac{(2j-1)!}{j!(j-1)!}.$$

Similarly,

$$\mathbb{P}\{M_{2j-3} = -1\} = \left(\frac{1}{2}\right)^{2j-3} \frac{(2j-3)!}{(j-1)!(j-2)!}.$$

It follows that, for $j \geq 2$,

$$\begin{aligned}\mathbb{P}\{\tau_1 = 2j-1\} &= \mathbb{P}\{\tau_1 \le 2j-1\} - \mathbb{P}\{\tau_1 \le 2j-3\}\\ &= \mathbb{P}\{M_{2j-3} = -1\} - \mathbb{P}\{M_{2j-1} = -1\}\\ &= \left(\frac{1}{2}\right)^{2j-3} \frac{(2j-3)!}{(j-1)!(j-2)!} - \left(\frac{1}{2}\right)^{2j-1} \frac{(2j-1)!}{j!(j-1)!}\\ &= \left(\frac{1}{2}\right)^{2j-1} \frac{(2j-3)!}{j!(j-1)!}\big[4j(j-1) - (2j-1)(2j-2)\big]\\ &= \left(\frac{1}{2}\right)^{2j-1} \frac{(2j-3)!}{j!(j-1)!}\big[2j(2j-2) - (2j-1)(2j-2)\big]\\ &= \left(\frac{1}{2}\right)^{2j-1} \frac{(2j-3)!}{j!(j-1)!}(2j-2)\\ &= \left(\frac{1}{2}\right)^{2j-1} \frac{(2j-2)!}{j!(j-1)!}.\end{aligned}$$

We have again obtained (5.2.22).

5.4 Perpetual American Put: An Example

In this section, we work out the pricing and hedging for a particular example of a *perpetual American put.* "Perpetual" refers to the fact that this put has no expiration date. This is not a traded instrument but rather a mathematical concept that serves as a bridge between the discrete-time American pricing and hedging discussion of Chapter 4 and the continuous-time analysis for American derivative securities in Chapter 8 of Volume II.

We consider a binomial model with up factor $u = 2$, down factor $d = \frac{1}{2}$, and interest rate $r = \frac{1}{4}$. For this model, the risk-neutral probabilities (see (1.1.8) of Chapter 1) are $\tilde{p} = \tilde{q} = \frac{1}{2}$. The stock price at time n is

$$S_n = S_0 \cdot 2^{M_n}, \tag{5.4.1}$$

where M_n is the random walk of (5.1.2). Under the risk-neutral probabilities, M_n is symmetric.

Consider an American put with strike price $K = 4$ and no expiration date. At any time n, the owner of this put can exercise it, selling for \$4 a share of stock worth $\$S_n$. We are interested in the value of this put as a function of the underlying stock price. Because there is no expiration, it is reasonable to expect the value of the put to depend only on the stock price, not on time. Similarly, it is reasonable to expect the optimal exercise policy to depend only on the stock price, not on time.

For the moment, let us suppose that $S_0 = 4$. Here are some possible exercise policies:

Policy 0: Exercise immediately. This corresponds to the stopping time τ_0, the first time the random walk M_n reaches the level 0, which is $\tau_0 = 0$. The associated value of this exercise policy is $V^{(\tau_0)} = 0$.

Policy 1: Exercise the first time the stock price falls to the level 2. This time is τ_{-1}, the first time the random walk falls to the level -1. We denote by $V^{(\tau_{-1})}$ the value of this exercise policy and compute it below.
Policy 2: Exercise the first time the stock price falls to the level 1. This time is τ_{-2}, the first time the random walk falls to the level -2. We denote by $V^{(\tau_{-2})}$ the value of this exercise policy and compute it below.

Let m be a positive integer. The risk-neutral value of the put if the owner uses the exercise policy τ_{-m}, which exercises the first time the stock price falls to $4 \cdot 2^{-m}$, is

$$V^{(\tau_{-m})} = \widetilde{\mathbb{E}}\left[\left(\frac{1}{1+r}\right)^{\tau_{-m}} (K - S_{\tau_{-m}})\right] = 4(1-2^{-m})\widetilde{\mathbb{E}}\left[\left(\frac{4}{5}\right)^{\tau_{-m}}\right]. \quad (5.4.2)$$

This is the risk-neutral expected payoff of the option at the time of exercise, discounted from the exercise time back to time zero. Because M_n is a symmetric random walk under the risk-neutral probabilities, we can compute the right-hand side of (5.4.2) using Theorem 5.2.3. We take α in that theorem to be $\frac{4}{5}$, so that

$$\frac{1-\sqrt{1-\alpha^2}}{\alpha} = \frac{5}{4} \cdot \left(1 - \sqrt{1 - \left(\frac{4}{5}\right)^2}\right) = \frac{5}{4} \cdot \left(1 - \sqrt{\frac{9}{25}}\right) = \frac{5}{4} \cdot \frac{2}{5} = \frac{1}{2}.$$

Theorem (5.2.3) implies that

$$V^{(\tau_{-m})} = 4(1-2^{-m})\left(\frac{1}{2}\right)^m, \quad m = 1, 2, \ldots. \quad (5.4.3)$$

In particular,

$$V^{(\tau_{-1})} = 4\left(1 - \frac{1}{2}\right)\frac{1}{2} = 1,$$
$$V^{(\tau_{-2})} = 4\left(1 - \frac{1}{4}\right)\frac{1}{4} = \frac{3}{4},$$
$$V^{(\tau_{-3})} = 4\left(1 - \frac{1}{8}\right)\frac{1}{8} = \frac{7}{16}.$$

Based on the computations above, we guess that the optimal policy is to exercise the first time the stock price falls to 2. This appears to give the option the largest value, at least if $S_0 = 4$. It is reasonable to expect that the optimality of this policy does not depend on the initial stock price. In other words, we expect that, regardless of the initial stock price, one should exercise the option the first time the stock price falls to 2. If the initial stock price is 2 or less, one would then exercise immediately.

To confirm the optimality of the exercise policy described in the previous paragraph, we first determine the value of the option for different initial stock prices when we use the policy of exercising the first time the stock price is at or below 2. If the initial stock price is $S_0 = 2^j$ for some integer $j \leq 1$, then $S_0 \leq 2$ and we exercise immediately. The value of the option under this exercise policy is the intrinsic value

$$v(2^j) = 4 - 2^j, \quad j = 1, 0, -1, -2, \ldots. \tag{5.4.4}$$

Now suppose the initial stock price is $S_0 = 2^j$ for some integer $j \geq 2$. We use the policy of exercising the first time the stock price reaches 2, which requires the random walk to fall to $-(j-1)$. We obtain the following values for the initial stock prices $S_0 = 4$ and $S_0 = 8$:

$$v(4) = \widetilde{\mathbb{E}}\left[\left(\frac{1}{1+r}\right)^{\tau_{-1}} (K - S_{\tau_{-1}})\right] = 2\widetilde{\mathbb{E}}\left[\left(\frac{4}{5}\right)^{\tau_{-1}}\right] = 2 \cdot \left(\frac{1}{2}\right)^1 = 1,$$

$$v(8) = \widetilde{\mathbb{E}}\left[\left(\frac{1}{1+r}\right)^{\tau_{-2}} (K - S_{\tau_{-2}})\right] = 2\widetilde{\mathbb{E}}\left[\left(\frac{4}{5}\right)^{\tau_{-2}}\right] = 2 \cdot \left(\frac{1}{2}\right)^2 = \frac{1}{2}.$$

In general,

$$v(2^j) = \widetilde{\mathbb{E}}\left[\left(\frac{4}{5}\right)^{\tau_{-(j-1)}} (4 - S_{\tau_{-(j-1)}})\right] = 2 \cdot \left(\frac{1}{2}\right)^{j-1} = \frac{4}{2^j}, \quad j = 2, 3, 4, \ldots. \tag{5.4.5}$$

Notice that, when $j = 1$, formula (5.4.5) gives the same value as (5.4.4).

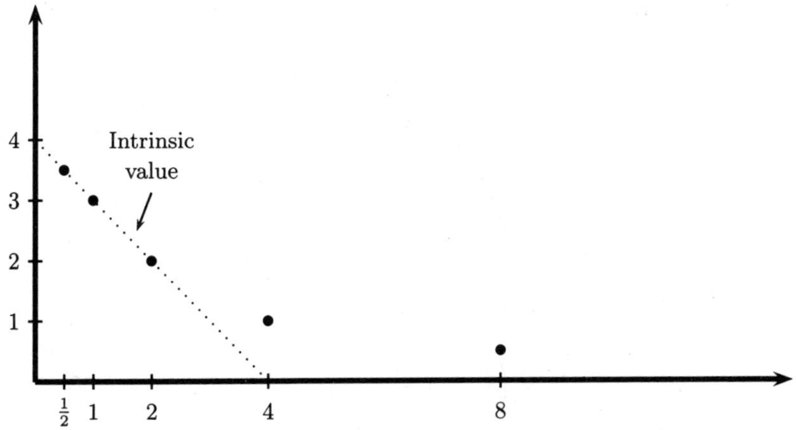

Fig. 5.4.1. Perpetual American put price $v(2^j)$.

We shall only consider stock prices of the form 2^j; if the initial stock price is of this form, then all subsequent stock prices are of this form (but with different integers j). At this point, we have a conjectured optimal exercise policy (exercise as soon as the stock price is 2 or below) and a conjectured option value function

$$v(2^j) = \begin{cases} 4 - 2^j, \text{ if } j \le 1, \\ \dfrac{4}{2^j}, \quad \text{if } j \ge 1. \end{cases} \tag{5.4.6}$$

To verify these conjectures, we need to establish analogues of the three properties in Theorem 4.4.2 of Chapter 4:

(i) $v(S_n) \ge (4 - S_n)^+$, $n = 0, 1, \ldots$;
(ii) the discounted process $\left(\frac{4}{5}\right)^n v(S_n)$ is a supermartingale under the risk-neutral probability measure;
(iii) $v(S_n)$ is the smallest process satisfying (i) and (ii).

Property (ii) guarantees that if we sell the option at time zero for $v(S_0)$, it is possible to take this initial capital and construct a hedge, sometimes consuming, so that the value of our hedging portfolio at each time n is $v(S_n)$. The details of this are given in the proof of Theorem 4.4.4 of Chapter 4. Property (i) guarantees that the value of the hedging portfolio, which is $v(S_n)$ at each time n, is sufficient to pay off the option when it is exercised. Taken together, properties (i) and (ii) guarantee that the seller is satisfied with this option price. Property (iii) guarantees that the buyer is also satisfied.

We now verify that properties (i), (ii), and (iii) are satisfied by the function $v(2^j)$ given by (5.4.6).

Property (i): For $j \le 1$ and $S_n = 2^j$, (5.4.6) implies immediately that $v(S_n) = 4 - S_n \ge (4 - S_n)^+$. For $j \ge 2$ and $S_n = 2^j$, we have $v(S_n) \ge 0 = (4 - S_n)^+$.

Property (ii): For $S_n = 2^j$ and $j \le 0$, we have

$$\begin{aligned}
\widetilde{\mathbb{E}}_n\left[\left(\frac{4}{5}\right)^{n+1} v(S_{n+1})\right] &= \frac{1}{2}\left(\frac{4}{5}\right)^{n+1} v(2^{j+1}) + \frac{1}{2}\left(\frac{4}{5}\right)^{n+1} v(2^{j-1}) \\
&= \left(\frac{4}{5}\right)^n \left[\frac{2}{5} v(2^{j+1}) + \frac{2}{5} v(2^{j-1})\right] \\
&= \left(\frac{4}{5}\right)^n \left[\frac{2}{5}(4 - 2^{j+1}) + \frac{2}{5}(4 - 2^{j-1})\right] \\
&= \left(\frac{4}{5}\right)^n \left[\frac{16}{5} - \frac{1}{5}(4+1)2^j\right] \\
&= \left(\frac{4}{5}\right)^n \left[\frac{16}{5} - 2^j\right] \\
&< \left(\frac{4}{5}\right)^n [4 - 2^j] = \left(\frac{4}{5}\right)^n v(S_n).
\end{aligned}$$

We thus see that in the "exercise region" of stock prices 1 and below, the discounted option value is a strict supermartingale. For $S_n = 2^j$ and $j \ge 2$, we have similarly

$$\widetilde{\mathbb{E}}\left[\left(\frac{4}{5}\right)^{n+1} v(S_{n+1})\right] = \frac{1}{2}\left(\frac{4}{5}\right)^{n+1} v(2^{j+1}) + \frac{1}{2}\left(\frac{4}{5}\right)^{n+1} v(2^{j-1})$$
$$= \left(\frac{4}{5}\right)^{n} \left[\frac{2}{5}v(2^{j+1}) + \frac{2}{5}v(2^{j-1})\right]$$
$$= \left(\frac{4}{5}\right)^{n} \left[\frac{2}{5}\cdot\frac{4}{2^{j+1}} + \frac{2}{5}\cdot\frac{4}{2^{j-1}}\right]$$
$$= \left(\frac{4}{5}\right)^{n} \left[\frac{4}{5\cdot 2^j} + \frac{16}{5\cdot 2^j}\right]$$
$$= \left(\frac{4}{5}\right)^{n} \cdot \frac{4}{2^j} = \left(\frac{4}{5}\right)^{n} v(S_n). \tag{5.4.7}$$

We thus see that in the "no exercise" region of stock prices 4 and above, the discounted option value is a martingale. It remains to consider the stock price $S_n = 2$ at the boundary between the "exercise" and "no exercise" regions. We compute

$$\widetilde{\mathbb{E}}\left[\left(\frac{4}{5}\right)^{n+1} v(S_{n+1})\right] = \frac{1}{2}\left(\frac{4}{5}\right)^{n+1} v(4) + \frac{1}{2}\left(\frac{4}{5}\right)^{n+1} v(1)$$
$$= \left(\frac{4}{5}\right)^{n} \left[\frac{2}{5}v(4) + \frac{2}{5}v(1)\right]$$
$$= \left(\frac{4}{5}\right)^{n} \left[\frac{2}{5}\cdot 1 + \frac{2}{5}\cdot 3\right]$$
$$< \left(\frac{4}{5}\right)^{n} \cdot 2 = \left(\frac{4}{5}\right)^{n} v(S_n).$$

Again, we have a strict supermartingale. This means that if the stock price is 2, the value derived from immediate exercise is strictly greater than the value derived from the right to future exercise. The option should be exercised when the stock price is 2. The same is true when the stock price is less than 2.

Property (iii): Let Y_n, $n = 1, 2, \ldots$ be a process that satisfies

(a) $Y_n \geq (4 - S_n)^+$, $n = 0, 1, \ldots$;
(b) the discounted process $\left(\frac{4}{5}\right)^n Y_n$ is a supermartingale under the risk-neutral probability measure.

We must show that $v(S_n) \leq Y_n$ for all n. To do this, we fix n and consider two cases. If $S_n \leq 2$, then (5.4.6) and (a) imply that $v(S_n) = 4 - S_n \leq Y_n$. In the other case, $S_n = 2^j$ for some $j \geq 2$. In this case, we let τ denote the first time after time n that the stock price falls to the level 2. Using (5.4.5) but starting at time n at position S_n rather than at time 0, we have

$$v(S_n) = \widetilde{\mathbb{E}}_n\left[\left(\frac{4}{5}\right)^{\tau-n} (4 - S_\tau)\right] = \widetilde{\mathbb{E}}_n\left[\left(\frac{4}{5}\right)^{\tau-n} (4 - S_\tau)^+\right], \tag{5.4.8}$$

where the second equality follows from the fact that $4 - S_\tau = 2 > 0$. On the other hand, from (b), the Optional Sampling Theorem 4.3.2 of Chapter 4, and (a), we know that, for all $k \geq n$,

$$\begin{aligned}\left(\frac{4}{5}\right)^n Y_n &= \left(\frac{4}{5}\right)^{\tau\wedge n} Y_{\tau\wedge n}\\ &\geq \widetilde{\mathbb{E}}_n\left[\left(\frac{4}{5}\right)^{\tau\wedge k} Y_{\tau\wedge k}\right]\\ &\geq \widetilde{\mathbb{E}}_n\left[\left(\frac{4}{5}\right)^{\tau\wedge k}(4 - S_{\tau\wedge k})^+\right].\end{aligned} \tag{5.4.9}$$

Letting $k \to \infty$ in (5.4.9), we see that

$$\left(\frac{4}{5}\right)^n Y_n \geq \widetilde{\mathbb{E}}_n\left[\left(\frac{4}{5}\right)^{\tau}(4 - S_\tau)^+\right].$$

Dividing by $\left(\frac{4}{5}\right)^n$, we obtain

$$Y_n \geq \widetilde{\mathbb{E}}_n\left[\left(\frac{4}{5}\right)^{\tau-n}(4 - S_\tau)^+\right]. \tag{5.4.10}$$

Relations (5.4.8) and (5.4.10) imply that $v(S_n) \leq Y_n$.

Property (iii) guarantees that the buyer of the American perpetual put is not being overcharged if she pays $v(S_n)$ at time n when the stock price is S_n. The key equality in this proof when $S_n \geq 4$ is (5.4.8), which says that if the buyer uses the policy of exercising the first time the stock price falls to the level 2, the risk-neutral expected discounted payoff is exactly what she is paying for the put. Of course, if $S_n \leq 2$ and the buyer exercises immediately, she recovers the purchase price $v(S_n)$ of the put, which is the intrinsic value for these stock prices. Whenever we have a conjectured price for an American derivative security that is derived by evaluating some exercise policy, the buyer is not being overcharged (because she can always use this policy) and property (iii) will be satisfied.

The proof we have just given of the optimality of the policy of exercising the first time the stock price is at or below 2 is probabilistic in nature. It considers the stochastic process $v(S_n)$ and shows that, when discounted, this process is a supermartingale under the risk-neutral probabilities. There is a second method of proof, which is the discrete-time version of the partial differential equation characterization of the value function v for the perpetual American put. We give that now.

First note that we may rewrite (5.4.6) as

$$v(s) = \begin{cases} 4 - s & \text{if } s \leq 2,\\ \dfrac{4}{s} & \text{if } s \geq 4,\end{cases} \tag{5.4.11}$$

where, as before, we are really only interested in values of s of the form $s = 2^j$ for some integer j. We may recast conditions (i), (ii), and (iii) as

(i)′ $v(s) \geq (4-s)^+$;

(ii)′ $v(s) \geq \frac{4}{5}\left[\frac{1}{2}v(2s) + \frac{1}{2}v\left(\frac{s}{2}\right)\right]$;

(iii)′ $v(s)$ is the smallest function satisfying (i)′ and (ii)′. In other words, if $w(s)$ is another function satisfying (i)′ and (ii)′, then $v(s) \leq w(s)$ for every s of the form $s = 2^j$.

It is clear that (i)′ and (iii)′ are just restatements of (i) and (iii). Property (ii)′ implies

$$\widetilde{\mathbb{E}}_n\left[\left(\frac{4}{5}\right)^{n+1} v(S_{n+1})\right] = \left(\frac{4}{5}\right)^n \frac{4}{5}\left[\frac{1}{2}v(2S_n) + \frac{1}{2}v\left(\frac{1}{2}S_n\right)\right] \leq \left(\frac{4}{5}\right)^n v(S_n),$$

which gives us property (ii).

If we have strict inequality in both (i)′ and (ii)′, then it is reasonable to expect that the function $v(s)$ could be made smaller and still satisfy (i)′ and (ii)′. This is indeed the case, although the proof is lengthy and will not be given. Because of property (iii)′, we cannot have strict inequality in both (i)′ and (ii)′. This gives us a fourth property:

(iv)′ For every s of the form $s = 2^j$, there is equality in either (i)′ or (ii)′.

In the case of $v(s)$ defined by (5.4.11), we have equality in (i)′ for $s \leq 2$. For $s \geq 4$, one can check, essentially as we did in (5.4.7), that equality holds in (ii)′:

$$\begin{aligned}\frac{4}{5}\left[\frac{1}{2}v(2s) + \frac{1}{2}v\left(\frac{s}{2}\right)\right] &= \frac{2}{5}\cdot\frac{4}{2s} + \frac{2}{5}\cdot\frac{8}{s}\\ &= \frac{4}{5s} + \frac{16}{5s} = \frac{20}{5s} = \frac{4}{s} = v(s).\end{aligned}$$

Properties (i)′, (ii)′, and (iv)′ are conveniently summarized by the single equation

$$v(s) = \max\left\{(4-s)^+, \frac{4}{5}\left[\frac{1}{2}v(2s) + \frac{1}{2}v\left(\frac{s}{2}\right)\right]\right\}. \tag{5.4.12}$$

But for values of $s > 4$, which make $4-s$ negative, $v(s)$ is equal to $\frac{4}{5}\left[\frac{1}{2}v(s) + \frac{1}{2}v\left(\frac{s}{2}\right)\right]$, not $(4-s)^+$. Therefore, for these values of s, it does not matter whether we write $(4-s)^+$ or $4-s$ in (5.4.12). For $s \leq 4$, the quantities $(4-s)^+$ and $4-s$ are the same. We may thus rewrite (5.4.12) more simply as

$$v(s) = \max\left\{4-s, \frac{4}{5}\left[\frac{1}{2}v(2s) + \frac{1}{2}v\left(\frac{s}{2}\right)\right]\right\}. \tag{5.4.13}$$

This is the so-called *Bellman equation* for the perpetual American put-pricing problem of this section.

The value $v(s)$ for the perpetual American put must satisfy (5.4.13), and this equation can be used to help determine the value. Unfortunately, there can be functions other than the one given by (5.4.11) that satisfy this equation. In particular, the function

$$w(s) = \frac{4}{s} \text{ for all } s = 2^j \tag{5.4.14}$$

also satisfies (5.4.13). The function $v(s)$ we seek is the *smallest* solution to (5.4.13) (see (iii)′). When we use (5.4.13) to solve for $v(s)$, we can use *boundary conditions* to rule out some of the extraneous solutions. In particular, the value of the perpetual American put must satisfy

$$\lim_{s \downarrow 0} v(s) = 4, \quad \lim_{s \to \infty} v(s) = 0. \tag{5.4.15}$$

The first of these conditions rules out the function $w(s)$ of (5.4.14).

In general, for a perpetual derivative security with intrinsic value $g(s)$, the value $v(s)$ of the option when the underlying stock price is s satisfies the Bellman equation

$$v(s) = \max\left\{g(s), \frac{1}{1+r}\left[\tilde{p}v(us) + \tilde{q}v(ds)\right]\right\}. \tag{5.4.16}$$

This is the same as equation (4.2.6) of Chapter 4 for the price of an American derivative security, except that when the derivative security is perpetual, its price does not depend on time; i.e., we have simply $v(s)$ rather than $v_n(s)$ and $v_{n+1}(s)$. If $g(s) = K - s$, so we are pricing a put with strike price K, then $v(s)$ should also satisfy the boundary conditions

$$\lim_{s \downarrow 0} v(s) = K, \quad \lim_{s \to \infty} v(s) = 0. \tag{5.4.17}$$

If $g(s) = s - K$, so we are pricing a call with strike price K, then $v(s)$ should satisfy the boundary conditions

$$\lim_{s \downarrow 0} v(s) = 0, \quad \lim_{s \to \infty} \frac{v(s)}{s} = 1. \tag{5.4.18}$$

(See Exercise 5.8.)

5.5 Summary

Let τ_m be the first time a symmetric random walk reaches level m, where m is a nonzero integer. Then

$$\mathbb{P}\{\tau_m < \infty\} = 1, \tag{5.2.12}$$

but

$$\mathbb{E}\tau_m = \infty. \tag{5.2.17}$$

If the random walk is asymmetric, with the probability p of an up step larger than $\frac{1}{2}$, then for m a positive integer, we have $\mathbb{P}\{\tau_m < \infty\} = 1$ and also $\mathbb{E}\tau_m < \infty$ (see Exercise 5.2 for the case of τ_1). If $0 < p < \frac{1}{2}$ and m is a positive integer, then $\mathbb{P}\{\tau_m = \infty\} > 0$ and $\mathbb{E}\tau_m = \infty$ (see Exercise 5.3 for the case of τ_1).

For both symmetric and asymmetric random walks, the moment generating function of τ_m can be computed (see (5.2.13) for the symmetric random walk, and see Exercises 5.2 and 5.3 for the asymmetric case). Knowledge of the moment-generating function permits us to determine the distribution of τ_m, although the computations are messy for large values of m. The distribution of τ_1 for the random walk with up-step probability p and down-step probability $q = 1 - p$ is

$$\mathbb{P}\{\tau_1 = 2j - 1\} = \frac{(2j-2)!}{j!(j-1)!} p^j q^{2j-1}, \quad j = 1, 2, \ldots. \tag{5.2.23}$$

See Exercise 5.4 for the case $m = 2$.

An alternative way to compute the distribution of τ_m is to use the *reflection principle*. The idea behind this principle is that there are as many paths that reach the level m prior to a final time but are below m at the final time as there are paths that are above m at the final time. This permits one to count the paths that reach level m by the final time. The reflection principle can also be used to determine the joint distribution of a random walk at some time n and the maximum value it achieves at or before time n; see Exercise 5.5.

The moment-generating function formula for a random walk permits us to evaluate the risk-neutral expected discounted payoff of a perpetual American put if it is exercised the first time the stock price falls to a specified threshold. One can then determine the exercise threshold that maximizes the risk-neutral expected discounted payoff. An example of this is worked out in Section 5.4. The argument that the put price found this way is indeed correct relies on verifying the three properties of the price of an American derivative security: (i) the price dominates the intrinsic value, (ii) the discounted price is a supermartingale under the risk-neutral measure, and (iii) the price obtained is the smallest possible price that satisfies (i) and (ii). These steps are carried out for the example in Section 5.4.

One can also seek the price of a perpetual American option by solving the Bellman equation, which for the put in Section 5.4 is

$$v(s) = \max\left\{4 - s, \frac{4}{5}\left[\frac{1}{2}v(2s) + \frac{1}{2}v\left(\frac{s}{2}\right)\right]\right\}. \tag{5.4.13}$$

This equation must hold for all values of s that are on the lattice of possible stock prices. The continuous-time version of this equation studied in Chapter 8 of Volume II must hold for all positive values of s. In both discrete and continuous time, the Bellman equation can have extraneous solutions. In continuous time, one can use boundary conditions appropriate for the option under consideration to rule these out. In discrete time, one must use both the boundary conditions and a careful consideration of the nature of the lattice of possible stock prices to rule out the extraneous solutions. In particular, the continuous-time version of this equation for the general perpetual American put problem has a simpler solution than the discrete-time version.

5.6 Notes

The solution to the perpetual American put problem considered in Section 5.4 was worked out as a 1994 Carnegie Mellon Summer Undergraduate Mathematics Institute project by Irene Villegas using the method outlined in Exercise 5.9. This is an example of an optimal stopping problem. A classical reference for optimal stopping problems is Shiryaev [39]. Shiryaev, Kabanov, Kramkov, and Melnikov [41] consider the general perpetual American put problem in a binomial model, and their work is reported in Shiryaev [40].

5.7 Exercises

Exercise 5.1. For the symmetric random walk, consider the first passage time τ_m to the level m. The random variable $\tau_2 - \tau_1$ is the number of steps required for the random walk to rise from level 1 to level 2, and this random variable has the same distribution as τ_1, the number of steps required for the random walk to rise from level 0 to level 1. Furthermore, $\tau_2 - \tau_1$ and τ_1 are independent of one another; the latter depends only on the coin tosses $1, 2, \dots, \tau_1$, and the former depends only on the coin tosses $\tau_1 + 1, \tau_1 + 2, \dots, \tau_2$.

(i) Use these facts to explain why

$$\mathbb{E}\alpha^{\tau_2} = (\mathbb{E}\alpha^{\tau_1})^2 \text{ for all } \alpha \in (0, 1).$$

(ii) Without using (5.2.13), explain why for any positive integer m we must have

$$\mathbb{E}\alpha^{\tau_m} = (\mathbb{E}\alpha^{\tau_1})^m \text{ for all } \alpha \in (0, 1). \tag{5.7.1}$$

(iii) Would equation (5.7.1) still hold if the random walk is not symmetric? Explain why or why not.

Exercise 5.2 (First passage time for random walk with upward drift). Consider the asymmetric random walk with probability p for an up step and probability $q = 1 - p$ for a down step, where $\frac{1}{2} < p < 1$ so that $0 < q < \frac{1}{2}$. In the notation of (5.2.1), let τ_1 be the first time the random walk starting from level 0 reaches level 1. If the random walk never reaches this level, then $\tau_1 = \infty$.

(i) Define $f(\sigma) = pe^{\sigma} + qe^{-\sigma}$. Show that $f(\sigma) > 1$ for all $\sigma > 0$.
(ii) Show that, when $\sigma > 0$, the process

$$S_n = e^{\sigma M_n}\left(\frac{1}{f(\sigma)}\right)^n$$

is a martingale.
(iii) Show that, for $\sigma > 0$,

$$e^{-\sigma} = \mathbb{E}\left[\mathbb{I}_{\{\tau_1<\infty\}}\left(\frac{1}{f(\sigma)}\right)^{\tau_1}\right].$$

Conclude that $\mathbb{P}\{\tau_1 < \infty\} = 1$.
(iv) Compute $\mathbb{E}\alpha^{\tau_1}$ for $\alpha \in (0, 1)$.
(v) Compute $\mathbb{E}\tau_1$.

Exercise 5.3 (First passage time for random walk with downward drift). Modify Exercise 5.2 by assuming $0 < p < \frac{1}{2}$ so that $\frac{1}{2} < q < 1$.

(i) Find a positive number σ_0 such that the function $f(\sigma) = pe^{\sigma} + qe^{-\sigma}$ satisfies $f(\sigma_0) = 1$ and $f(\sigma) > 1$ for all $\sigma > \sigma_0$.
(ii) Determine $\mathbb{P}\{\tau_1 < \infty\}$. (This quantity is no longer equal to 1.)
(iii) Compute $\mathbb{E}\alpha^{\tau_1}$ for $\alpha \in (0, 1)$.
(iv) Compute $\mathbb{E}\left[\mathbb{I}_{\{\tau_1<\infty\}}\tau_1\right]$. (Since $\mathbb{P}\{\tau_1 = \infty\} > 0$, we have $\mathbb{E}\tau_1 = \infty$.)

Exercise 5.4 (Distribution of τ_2). Consider the symmetric random walk, and let τ_2 be the first time the random walk, starting from level 0, reaches the level 2. According to Theorem 5.2.3,

$$\mathbb{E}\alpha^{\tau_2} = \left(\frac{1-\sqrt{1-\alpha^2}}{\alpha}\right)^2 \quad \text{for all } \alpha \in (0, 1).$$

Using the power series (5.2.21), we may write the right-hand side as

$$\begin{aligned}\left(\frac{1-\sqrt{1-\alpha^2}}{\alpha}\right)^2 &= \frac{2}{\alpha}\cdot\frac{1-\sqrt{1-\alpha^2}}{\alpha} - 1\\ &= -1 + \sum_{j=1}^{\infty}\left(\frac{\alpha}{2}\right)^{2j-2}\frac{(2j-2)!}{j!(j-1)!}\\ &= \sum_{j=2}^{\infty}\left(\frac{\alpha}{2}\right)^{2j-2}\frac{(2j-2)!}{j!(j-1)!}\\ &= \sum_{k=1}^{\infty}\left(\frac{\alpha}{2}\right)^{2k}\frac{(2k)!}{(k+1)!k!}.\end{aligned}$$

(i) Use the power series above to determine $\mathbb{P}\{\tau_2 = 2k\}$, $k = 1, 2, \ldots$.
(ii) Use the reflection principle to determine $\mathbb{P}\{\tau_2 = 2k\}$, $k = 1, 2, \ldots$.

Exercise 5.5 (Joint distribution of random walk and maximum-to-date). Let M_n be a symmetric random walk, and define its *maximum-to-date* process

$$M_n^* = \max_{1 \le k \le n} M_k. \tag{5.7.2}$$

Let n and m be even positive integers, and let b be an even integer less than or equal to m. Assume $m \le n$ and $2m - b \le n$.

(i) Use an argument based on reflected paths to show that

$$\begin{aligned}\mathbb{P}\{M_n^* \ge m, M_n = b\} &= \mathbb{P}\{M_n = 2m - b\} \\ &= \frac{n!}{\left(\frac{n-b}{2} + m\right)! \left(\frac{n+b}{2} - m\right)!} \left(\frac{1}{2}\right)^n.\end{aligned}$$

(ii) If the random walk is asymmetric with probability p for an up step and probability $q = 1 - p$ for a down step, where $0 < p < 1$, what is $\mathbb{P}\{M_n^* \ge m, M_n = b\}$?

Exercise 5.6. The value of the perpetual American put in Section 5.4 is the limit as $n \to \infty$ of the value of an American put with the same strike price 4 that expires at time n. When the initial stock price is $S_0 = 4$, the value of the perpetual American put is 1 (see (5.4.6) with $j = 2$). Show that the value of an American put in the same model when the initial stock price is $S_0 = 4$ is 0.80 if the put expires at time 1, 0.928 if the put expires at time 3, and 0.96896 if the put expires at time 5.

Exercise 5.7 (Hedging a short position in the perpetual American put). Suppose you have sold the perpetual American put of Section 5.4 and are hedging the short position in this put. Suppose that at the current time the stock price is s and the value of your hedging portfolio is $v(s)$. Your hedge is to first consume the amount

$$c(s) = v(s) - \frac{4}{5}\left[\frac{1}{2}v(2s) + \frac{1}{2}v\left(\frac{s}{2}\right)\right] \tag{5.7.3}$$

and then take a position

$$\delta(s) = \frac{v(2s) - v\left(\frac{s}{2}\right)}{2s - \frac{s}{2}} \tag{5.7.4}$$

in the stock. (See Theorem 4.2.2 of Chapter 4. The processes C_n and Δ_n in that theorem are obtained by replacing the dummy variable s by the stock price S_n in (5.7.3) and (5.7.4); i.e., $C_n = c(S_n)$ and $\Delta_n = \delta(S_n)$.) If you hedge this way, then regardless of whether the stock goes up or down on the next step, the value of your hedging portfolio should agree with the value of the perpetual American put.

(i) Compute $c(s)$ when $s = 2^j$ for the three cases $j \leq 0$, $j = 1$, and $j \geq 2$.
(ii) Compute $\delta(s)$ when $s = 2^j$ for the three cases $j \leq 0$, $j = 1$, and $j \geq 2$.
(iii) Verify in each of the three cases $s = 2^j$ for $j \leq 0$, $j = 1$, and $j \geq 2$ that the hedge works (i.e., regardless of whether the stock goes up or down, the value of your hedging portfolio at the next time is equal to the value of the perpetual American put at that time).

Exercise 5.8 (Perpetual American call). Like the perpetual American put of Section 5.4, the perpetual American call has no expiration. Consider a binomial model with up factor u, down factor d, and interest rate $r \geq 0$ that satisfies the no-arbitrage condition $0 < d < 1 + r < u$. The risk-neutral probabilities are

$$\tilde{p} = \frac{1+r-d}{u-d}, \quad \tilde{q} = \frac{u-1-r}{u-d}.$$

The intrinsic value of the perpetual American call is $g(s) = s - K$, where $K > 0$ is the strike price. The purpose of this exercise is to show that the value of the call is always the price of the underlying stock, and there is no optimal exercise time.

(i) Let $v(s) = s$. Show that $v(S_n)$ is always at least as large as the intrinsic value $g(S_n)$ of the call and $\left(\frac{1}{1+r}\right)^n v(S_n)$ is a supermartingale under the risk-neutral probabilities. These are the analogues of properties (i) and (ii) for the perpetual American put of Section 5.4.
(ii) To show that $v(s) = s$ is not too large to be the value of the perpetual American call, we must find a good policy for the purchaser of the call. Show that if the purchaser of the call exercises at time n, regardless of the stock price at that time, then the discounted risk-neutral expectation of her payoff is $S_0 - \frac{K}{(1+r)^n}$. Because this is true for every n, and

$$\lim_{n\to\infty} \left[S_0 - \frac{K}{(1+r)^n} \right] = S_0,$$

the value of the call at time zero must be at least S_0. (The same is true at all other times; the value of the call is at least as great as the current stock price.)
(iii) In place of (i) and (ii) above, we could verify that $v(s) = s$ is the value of the perpetual American call by checking that this function satisfies the equation (5.4.16) and boundary conditions (5.4.18). Do this verification.
(iv) Show that there is no optimal time to exercise the perpetual American call.

Exercise 5.9. (Provided by Irene Villegas.) Here is a method for solving equation (5.4.13) for the value of the perpetual American put in Section 5.4.

(i) We first determine $v(s)$ for large values of s. When s is large, it is not optimal to exercise the put, so the maximum in (5.4.13) will be given by the second term,

$$\frac{4}{5}\left[\frac{1}{2}v(2s)+\frac{1}{2}v\left(\frac{s}{2}\right)\right]=\frac{2}{5}v(2s)+\frac{2}{5}v\left(\frac{s}{2}\right).$$

We thus seek solutions to the equation

$$v(s)=\frac{2}{5}v(2s)+\frac{2}{5}v\left(\frac{s}{2}\right). \tag{5.7.5}$$

All such solutions are of the form s^p for some constant p or linear combinations of functions of this form. Substitute s^p into (5.7.5), obtain a quadratic equation for 2^p, and solve to obtain $2^p = 2$ or $2^p = \frac{1}{2}$. This leads to the values $p = 1$ and $p = -1$, i.e., $v_1(s) = s$ and $v_2(s) = \frac{1}{s}$ are solutions to (5.7.5).

(ii) The general solution to (5.7.5) is a linear combination of $v_1(s)$ and $v_2(s)$, i.e.,

$$v(s)=As+\frac{B}{s}. \tag{5.7.6}$$

For large values of s, the value of the perpetual American put must be given by (5.7.6). It remains to evaluate A and B. Using the second boundary condition in (5.4.15), show that A must be zero.

(iii) We have thus established that for large values of s, $v(s) = \frac{B}{s}$ for some constant B still to be determined. For small values of s, the value of the put is its intrinsic value $4 - s$. We must choose B so these two functions coincide at some point, i.e., we must find a value for B so that, for some $s > 0$,

$$f_B(s)=\frac{B}{s}-(4-s)$$

equals zero. Show that, when $B > 4$, this function does not take the value 0 for any $s > 0$, but, when $B \leq 4$, the equation $f_B(s) = 0$ has a solution.

(iv) Let B be less than or equal to 4, and let s_B be a solution of the equation $f_B(s) = 0$. Suppose s_B is a stock price that can be attained in the model (i.e., $s_B = 2^j$ for some integer j). Suppose further that the owner of the perpetual American put exercises the first time the stock price is s_B or smaller. Then the discounted risk-neutral expected payoff of the put is $v_B(S_0)$, where $v_B(s)$ is given by the formula

$$v_B(s)=\begin{cases}4-s, & \text{if } s \leq s_B,\\ \frac{B}{s}, & \text{if } s \geq s_B.\end{cases} \tag{5.7.7}$$

Which values of B and s_B give the owner the largest option value?

(v) For $s < s_B$, the derivative of $v_B(s)$ is $v_B'(s) = -1$. For $s > s_B$, this derivative is $v_B'(s) = -\frac{B}{s^2}$. Show that the best value of B for the option owner makes the derivative of $v_B(s)$ continuous at $s = s_B$ (i.e., the two formulas for $v_B'(s)$ give the same answer at $s = s_B$).

6

Interest-Rate-Dependent Assets

6.1 Introduction

In this chapter, we develop a simple, binomial model for interest rates and then examine some common assets whose value depends on interest rates. Assets in this class are called *fixed income* assets.

The simplest fixed income asset is a *zero-coupon bond*, a bond that pays a specified amount (called its *face value* or *par value*) at a specified time (called *maturity*). At times prior to maturity, the value of this asset is less than its face value, provided the interest rate is always greater than zero. One defines the *yield* of the zero-coupon bond corresponding to a given maturity as the constant interest rate that would be needed so that the time-zero price of the bond invested at time zero and allowed to accumulate at this interest rate would grow to the face value of the bond at the bond's maturity. Because there is theoretically a zero-coupon bond for each possible maturity, there is theoretically a yield corresponding to every time greater than the current time, which we call 0. The *yield curve* is the function from the maturity variable to the yield variable. One is interested in building models that not only determine the yield curve at a particular time but provide a method for random evolution of the yield curve forward in time. These are called *term structure of interest rates* models.

The existence of a zero-coupon bond for each maturity raises the specter of arbitrage. A term-structure model has numerous tradable assets—all the zero-coupon bonds and perhaps other fixed income assets. How can one be assured that one cannot within the model find an arbitrage by trading in these instruments? Regardless of one's view about the existence of arbitrage in the real world, it is clear that models built for pricing and hedging must not admit arbitrage. Any question that one attempts to answer with a model permitting arbitrage has a nonsensical answer. What is the price of an option on an asset? If one can begin with zero initial capital and hedge a short position in the option by using the arbitrage possibility within the model, one could argue that the option price is zero. Furthermore, one cannot advance a more convincing argument for any other price.

In this chapter, we avoid the problem of arbitrage by building the model under the risk-neutral measure. In other words, we first describe the evolution of the interest rate under the risk-neutral measure and then determine the prices of zero-coupon bonds and all other fixed income assets by using the risk-neutral pricing formula. This construction guarantees that all discounted asset prices are martingales, and so all discounted portfolio processes are martingales (when all coupons and other payouts of cash are reinvested in the portfolio). This is Theorem 6.2.6, the main result of Section 6.2. Because a martingale that begins at zero must always have expectation zero, martingale discounted portfolio processes are inconsistent with arbitrage; they cannot have a positive value with positive probability at some time unless they also have a negative value with positive probability at that time.

Models that are built starting from the interest rate and using risk-neutral pricing are called *short-rate models.* Examples of such models in continuous time are the Vasicek-Hull-White model [42], [23], and the Cox-Ingersoll-Ross model [10]. The classical model in continuous time that begins with a description of the random evolution of the yield curve and develops conditions under which the model is arbitrage-free is due to Heath, Jarrow, and Morton [20], [21]. This model might be called a *whole yield model* because it takes the whole yield curve as a starting point. The difference between these two classes of models is of fundamental practical importance. However, in a whole yield model, one still has an interest rate, and all the other assets are related to the interest rate by the risk-neutral pricing formula. Although the interest rate evolution in whole yield models is generally more complex than in the binomial model of this chapter, the ideas we develop here can be transferred to whole yield models.

Two common discrete-time models are Ho and Lee [22] and Black-Derman-Toy [4]. We use the former as the basis for Example 6.4.4 and the latter as the basis for Example 6.5.5.

6.2 Binomial Model for Interest Rates

Let Ω be the set of 2^N possible outcomes $\omega_1\omega_2\dots\omega_N$ of N tosses of a coin, and let $\widetilde{\mathbb{P}}$ be a probability measure on Ω under which every sequence $\omega_1\omega_2\dots\omega_N$ has strictly positive probability. We define an *interest rate process* to be a sequence of random variables

$$R_0, R_1, \dots, R_{N-1},$$

where R_0 is not random and, for $n = 1, \dots, N-1$, R_n depends only on the first n coin tosses $\omega_1 \dots \omega_n$. One dollar invested in the money market at time n grows to $1 + R_n$ dollars at time $n+1$. Although the interest rate is random, at time n we know the interest rate that will be applied to money market investments over the period from time n to time $n+1$. This is less random than a stock. If we invest in a stock at time n, we do not know what the value of the investment will be at time $n+1$.

It is natural to assume that $R_n > 0$ for all n and all $\omega_1 \dots \omega_n$. This is the case to keep in mind. However, the analysis requires only that

$$R_n(\omega_1 \dots \omega_n) > -1 \text{ for all } n \text{ and all } \omega_1, \dots, \omega_n, \tag{6.2.1}$$

and this is the only assumption we make.

We define the *discount process* by

$$D_n = \frac{1}{(1+R_0)\cdots(1+R_{n-1})}, \quad n = 1, 2, \dots, N; \quad D_0 = 1. \tag{6.2.2}$$

Note that D_n depends on only the first $n-1$ coin tosses, contrary to the usual situation in which the subscript indicates the number of coin tosses on which the random variable depends.

The risk-neutral pricing formula says that the value at time zero of a payment X received at time m (where X is allowed to depend only on $\omega_1 \dots \omega_m$) is

$$\widetilde{\mathbb{E}}[D_m X],$$

the risk-neutral expected discounted payment. We use this formula to define the time-zero price of a zero-coupon bond that pays 1 at maturity time m to be

$$B_{0,m} = \widetilde{\mathbb{E}}[D_m]. \tag{6.2.3}$$

The yield for this bond is the number y_m for which

$$\frac{1}{B_{0,m}} = (1+y_m)^m.$$

At time zero, an investment of one dollar in the m-maturity bond would purchase $\frac{1}{B_{0,m}}$ of these bonds, and this investment would pay off $\frac{1}{B_{0,m}}$ at time m. This is the same as investing one dollar at a constant rate of interest y_m between times 0 and m. Of course, we can solve the equation above for y_m:

$$y_m = \left(\frac{1}{B_{0,m}}\right)^{\frac{1}{m}} - 1.$$

We would also like to define the price of the m-maturity zero-coupon bond at time n, where $1 \le n \le m$. To do that, we need to extend the notion of conditional expectation of Definition 2.3.1. Consider the following example.

Example 6.2.1. Assume $N = 3$ so that

$$\Omega = \{HHH, HHT, HTH, HTT, THH, THT, TTH, TTT\}.$$

Assume further that

$$\widetilde{\mathbb{P}}\{HHH\} = \frac{2}{9}, \quad \widetilde{\mathbb{P}}\{HHT\} = \frac{1}{9}, \quad \widetilde{\mathbb{P}}\{HTH\} = \frac{1}{12}, \quad \widetilde{\mathbb{P}}\{HTT\} = \frac{1}{12}$$
$$\widetilde{\mathbb{P}}\{THH\} = \frac{1}{6}, \quad \widetilde{\mathbb{P}}\{THT\} = \frac{1}{12}, \quad \widetilde{\mathbb{P}}\{TTH\} = \frac{1}{8}, \quad \widetilde{\mathbb{P}}\{TTT\} = \frac{1}{8}.$$

These numbers sum to 1. We define the sets

$$\begin{aligned} A_{HH} &= \{\omega_1 = H, \omega_2 = H\} = \{HHH, HHT\}, \\ A_{HT} &= \{\omega_1 = H, \omega_2 = T\} = \{HTH, HTT\}, \\ A_{TH} &= \{\omega_1 = T, \omega_2 = H\} = \{THH, THT\}, \\ A_{TT} &= \{\omega_1 = T, \omega_2 = T\} = \{TTH, TTT\} \end{aligned}$$

of outcomes that begin with HH, HT, TH, and TT on the first two tosses, respectively. We have

$$\widetilde{\mathbb{P}}\{A_{HH}\} = \frac{1}{3}, \quad \widetilde{\mathbb{P}}\{A_{HT}\} = \frac{1}{6}, \quad \widetilde{\mathbb{P}}\{A_{TH}\} = \frac{1}{4}, \quad \widetilde{\mathbb{P}}\{A_{TT}\} = \frac{1}{4}.$$

Similarly, we define the sets

$$\begin{aligned} A_H &= \{\omega_1 = H\} = \{HHH, HHT, HTH, HTT\}, \\ A_T &= \{\omega_1 = T\} = \{THH, THT, TTH, TTT\} \end{aligned}$$

of outcomes that begin with H and T on the first toss, respectively. For these sets, we have

$$\widetilde{\mathbb{P}}\{A_H\} = \frac{1}{2}, \quad \widetilde{\mathbb{P}}\{A_T\} = \frac{1}{2}.$$

If the coin is tossed three times and we are told only that the first two tosses are HH, then we know that the outcome of the three tosses is in A_{HH}. Conditioned on this information, the probability that the third toss is an H is

$$\widetilde{\mathbb{P}}\{\omega_3 = H | \omega_1 = H, \omega_2 = H\} = \frac{\widetilde{\mathbb{P}}\{HHH\}}{\widetilde{\mathbb{P}}\{A_{HH}\}} = \frac{2/9}{1/3} = \frac{2}{3}.$$

Similarly, we have

$$\begin{aligned} \widetilde{\mathbb{P}}\{\omega_3 = T | \omega_1 = H, \omega_2 = H\} &= \frac{\widetilde{\mathbb{P}}\{HHT\}}{\widetilde{\mathbb{P}}\{A_{HH}\}} = \frac{1/9}{1/3} = \frac{1}{3}, \\ \widetilde{\mathbb{P}}\{\omega_3 = H | \omega_1 = H, \omega_2 = T\} &= \frac{\widetilde{\mathbb{P}}\{HTH\}}{\widetilde{\mathbb{P}}\{A_{HT}\}} = \frac{1/12}{1/6} = \frac{1}{2}, \\ \widetilde{\mathbb{P}}\{\omega_3 = T | \omega_1 = H, \omega_2 = T\} &= \frac{\widetilde{\mathbb{P}}\{HTT\}}{\widetilde{\mathbb{P}}\{A_{HT}\}} = \frac{1/12}{1/6} = \frac{1}{2}, \end{aligned}$$

$$\widetilde{\mathbb{P}}\{\omega_3 = H|\omega_1 = T, \omega_2 = H\} = \frac{\widetilde{\mathbb{P}}\{THH\}}{\widetilde{\mathbb{P}}\{A_{TH}\}} = \frac{1/6}{1/4} = \frac{2}{3},$$

$$\widetilde{\mathbb{P}}\{\omega_3 = T|\omega_1 = T, \omega_2 = H\} = \frac{\widetilde{\mathbb{P}}\{THT\}}{\widetilde{\mathbb{P}}\{A_{TH}\}} = \frac{1/12}{1/4} = \frac{1}{3},$$

$$\widetilde{\mathbb{P}}\{\omega_3 = H|\omega_1 = T, \omega_2 = T\} = \frac{\widetilde{\mathbb{P}}\{TTH\}}{\widetilde{\mathbb{P}}\{A_{TT}\}} = \frac{1/8}{1/4} = \frac{1}{2},$$

$$\widetilde{\mathbb{P}}\{\omega_3 = T|\omega_1 = T, \omega_2 = T\} = \frac{\widetilde{\mathbb{P}}\{TTT\}}{\widetilde{\mathbb{P}}\{A_{TT}\}} = \frac{1/8}{1/4} = \frac{1}{2}.$$

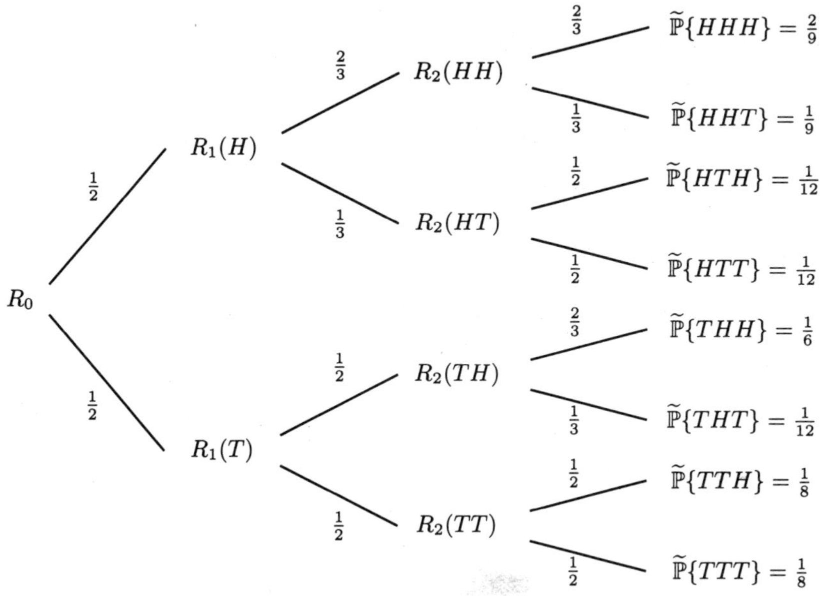

Fig. 6.2.1. A three-period interest rate model.

If the coin is tossed three times and we are told only that the first toss is H, then we know that the outcome of the three tosses is in A_H. Conditioned on this information, the probability that the second toss is H is

$$\widetilde{\mathbb{P}}\{\omega_2 = H|\omega_1 = H\} = \frac{\widetilde{\mathbb{P}}\{A_{HH}\}}{\widetilde{\mathbb{P}}\{A_H\}} = \frac{1/3}{1/2} = \frac{2}{3}.$$

Similarly,

$$\widetilde{\mathbb{P}}\{\omega_2 = T|\omega_1 = H\} = \frac{\widetilde{\mathbb{P}}\{A_{HT}\}}{\widetilde{\mathbb{P}}\{A_H\}} = \frac{1/6}{1/2} = \frac{1}{3},$$

$$\widetilde{\mathbb{P}}\{\omega_2 = H|\omega_1 = T\} = \frac{\widetilde{\mathbb{P}}\{A_{TH}\}}{\widetilde{\mathbb{P}}\{A_T\}} = \frac{1/4}{1/2} = \frac{1}{2},$$

$$\widetilde{\mathbb{P}}\{\omega_2 = T|\omega_1 = T\} = \frac{\widetilde{\mathbb{P}}\{A_{TT}\}}{\widetilde{\mathbb{P}}\{A_T\}} = \frac{1/4}{1/2} = \frac{1}{2}.$$

These conditional probabilities are the probabilities of traversing the indicated links in the tree shown in Figure 6.2.1. We shall call them the *transition probabilities* associated with $\widetilde{\mathbb{P}}$. The (unconditional) probabilities of following each of the eight paths are shown at the right ends of the paths.

We may compute $\widetilde{\mathbb{P}}\{\omega_2 = H, \omega_3 = H|\omega_1 = H\}$ by two methods, either computing along the upper links in Figure 6.2.1 using the formula

$$\begin{aligned}
&\widetilde{\mathbb{P}}\{\omega_2 = H, \omega_3 = H|\omega_1 = H\} \\
&\quad = \widetilde{\mathbb{P}}\{\omega_2 = H|\omega_1 = H\} \cdot \widetilde{\mathbb{P}}\{\omega_3 = H|\omega_1 = H, \omega_2 = H\} \\
&\quad = \frac{2}{3} \cdot \frac{2}{3} = \frac{4}{9}
\end{aligned}$$

or by using the formula

$$\widetilde{\mathbb{P}}\{\omega_2 = H, \omega_3 = H|\omega_1 = H\} = \frac{\widetilde{\mathbb{P}}\{HHH\}}{\widetilde{\mathbb{P}}(A_H)} = \frac{2/9}{1/2} = \frac{4}{9}.$$

By either of these two methods, we have the following formulas:

$$\widetilde{\mathbb{P}}\{\omega_2 = H, \omega_3 = H|\omega_1 = H\} = \frac{4}{9}, \quad \widetilde{\mathbb{P}}\{\omega_2 = H, \omega_3 = T|\omega_1 = H\} = \frac{2}{9},$$

$$\widetilde{\mathbb{P}}\{\omega_2 = T, \omega_3 = H|\omega_1 = H\} = \frac{1}{6}, \quad \widetilde{\mathbb{P}}\{\omega_2 = T, \omega_3 = T|\omega_1 = H\} = \frac{1}{6},$$

$$\widetilde{\mathbb{P}}\{\omega_2 = H, \omega_3 = H|\omega_1 = T\} = \frac{1}{3}, \quad \widetilde{\mathbb{P}}\{\omega_2 = H, \omega_3 = T|\omega_1 = T\} = \frac{1}{6},$$

$$\widetilde{\mathbb{P}}\{\omega_2 = T, \omega_3 = H|\omega_1 = T\} = \frac{1}{4}, \quad \widetilde{\mathbb{P}}\{\omega_2 = T, \omega_3 = T|\omega_1 = T\} = \frac{1}{4}.$$

Finally, suppose we have a random variable $X(\omega_1\omega_2\omega_3)$. We define the conditional expectation of this random variable, given the first coin toss, by the formulas

$$\begin{aligned}
\widetilde{\mathbb{E}}_1[X](H) = {} & X(HHH)\,\widetilde{\mathbb{P}}\{\omega_2 = H, \omega_3 = H|\omega_1 = H\} \\
& + X(HHT)\,\widetilde{\mathbb{P}}\{\omega_2 = H, \omega_3 = T|\omega_1 = H\} \\
& + X(HTH)\,\widetilde{\mathbb{P}}\{\omega_2 = T, \omega_3 = H|\omega_1 = H\} \\
& + X(HTT)\,\widetilde{\mathbb{P}}\{\omega_2 = T, \omega_3 = T|\omega_1 = H\} \\
= {} & \frac{4}{9}X(HHH) + \frac{2}{9}X(HHT) + \frac{1}{6}X(HTH) + \frac{1}{6}X(HTT),
\end{aligned}$$

$$\begin{aligned}
\widetilde{\mathbb{E}}_1[X](T) &= X(THH)\,\widetilde{\mathbb{P}}\{\omega_2 = H, \omega_3 = H | \omega_1 = T\} \\
&\quad + X(THT)\,\widetilde{\mathbb{P}}\{\omega_2 = H, \omega_3 = T | \omega_1 = T\} \\
&\quad + X(TTH)\,\widetilde{\mathbb{P}}\{\omega_2 = T, \omega_3 = H | \omega_1 = T\} \\
&\quad + X(TTT)\,\widetilde{\mathbb{P}}\{\omega_2 = T, \omega_3 = T | \omega_1 = T\} \\
&= \frac{1}{3}X(THH) + \frac{1}{6}X(THT) + \frac{1}{4}X(TTH) + \frac{1}{4}X(TTT).
\end{aligned}$$

Note that $\widetilde{\mathbb{E}}_1[X]$ is a random variable because it depends on the first toss. We define the conditional expectation of X, given the first two tosses, by the formulas

$$\begin{aligned}
\widetilde{\mathbb{E}}_2[X](HH) &= X(HHH)\,\widetilde{\mathbb{P}}\{\omega_3 = H | \omega_1 = H, \omega_2 = H\} \\
&\quad + X(HHT)\,\widetilde{\mathbb{P}}\{\omega_3 = T | \omega_1 = H, \omega_2 = H\} \\
&= \frac{2}{3}X(HHH) + \frac{1}{3}X(HHT), \\
\widetilde{\mathbb{E}}_2[X](HT) &= X(HTH)\,\widetilde{\mathbb{P}}\{\omega_3 = H | \omega_1 = H, \omega_2 = T\} \\
&\quad + X(HTT)\,\widetilde{\mathbb{P}}\{\omega_3 = T | \omega_1 = H, \omega_2 = T\} \\
&= \frac{1}{2}X(HTH) + \frac{1}{2}X(HTT), \\
\widetilde{\mathbb{E}}_2[X](TH) &= X(THH)\,\widetilde{\mathbb{P}}\{\omega_3 = H | \omega_1 = T, \omega_2 = H\} \\
&\quad + X(THT)\,\widetilde{\mathbb{P}}\{\omega_3 = T | \omega_1 = T, \omega_2 = H\} \\
&= \frac{2}{3}X(THH) + \frac{1}{3}X(THT), \\
\widetilde{\mathbb{E}}_2[X](TT) &= X(TTH)\,\widetilde{\mathbb{P}}\{\omega_3 = H | \omega_1 = T, \omega_2 = T\} \\
&\quad + X(TTT)\,\widetilde{\mathbb{P}}\{\omega_3 = T | \omega_1 = T, \omega_2 = T\} \\
&= \frac{1}{2}X(TTH) + \frac{1}{2}X(TTT).
\end{aligned}$$

Note that $\widetilde{\mathbb{E}}_2[X]$ is also a random variable; it depends on the first two tosses.

If the random variable X whose conditional expectation we are computing depends on only the first two tosses, then the computations above simplify. In particular, we would have

$$\begin{aligned}
\widetilde{\mathbb{E}}_1[X](H) &= \left(\frac{4}{9} + \frac{2}{9}\right) X(HH) + \left(\frac{1}{6} + \frac{1}{6}\right) X(HT) \\
&= \frac{2}{3}X(HH) + \frac{1}{3}X(HT),
\end{aligned}$$

$$\begin{aligned}
\widetilde{\mathbb{E}}_1[X](T) &= \left(\frac{1}{3} + \frac{1}{6}\right) X(TH) + \left(\frac{1}{4} + \frac{1}{4}\right) X(TT) \\
&= \frac{1}{2}X(TH) + \frac{1}{2}X(TT),
\end{aligned}$$

and

$$\widetilde{\mathbb{E}}_2[X](HH) = X(HH),\ \widetilde{\mathbb{E}}_2[X](HT) = X(HT),$$
$$\widetilde{\mathbb{E}}_2[X](TH) = X(TH),\ \widetilde{\mathbb{E}}_2[X](TT) = X(TT). \qquad \square$$

The previous example is a special case of the following definition.

Definition 6.2.2. *Let $\widetilde{\mathbb{P}}$ be a probability measure on the space Ω of all possible sequences of N coin tosses. Assume that every sequence $\omega_1 \dots \omega_N$ in Ω has positive probability under $\widetilde{\mathbb{P}}$. Let $1 \le n \le N-1$, and let $\overline{\omega}_1 \dots \overline{\omega}_N$ be a sequence of N coin tosses. We define*

$$\widetilde{\mathbb{P}}\{\omega_{n+1} = \overline{\omega}_{n+1}, \dots, \omega_N = \overline{\omega}_N | \omega_1 = \overline{\omega}_1, \dots, \omega_n = \overline{\omega}_n\} = \frac{\widetilde{\mathbb{P}}\{\overline{\omega}_1 \dots \overline{\omega}_n \overline{\omega}_{n+1} \dots \overline{\omega}_N\}}{\widetilde{\mathbb{P}}\{\omega_1 = \overline{\omega}_1, \dots, \omega_n = \overline{\omega}_n\}}.$$

Let X be a random variable. We define the conditional expectation of X based on the information at time n *by the formula*

$$\widetilde{\mathbb{E}}_n[X](\overline{\omega}_1 \dots \overline{\omega}_n) = \sum_{\overline{\omega}_{n+1}, \dots, \overline{\omega}_N} X(\overline{\omega}_1 \dots \overline{\omega}_n \overline{\omega}_{n+1} \dots \overline{\omega}_N) \times \widetilde{\mathbb{P}}\{\omega_{n+1} = \overline{\omega}_{n+1}, \dots, \omega_N = \overline{\omega}_N | \omega_1 = \overline{\omega}_1, \dots, \omega_n = \overline{\omega}_n\}.$$

We further set $\widetilde{\mathbb{E}}_0[X] = \widetilde{\mathbb{E}}X$ and $\widetilde{\mathbb{E}}_N[X] = X$.

Remark 6.2.3. Conditional expectation as defined above is a generalization of Definition 2.3.1 that allows for nonindependent coin tosses. It satisfies all the properties of Theorem 2.3.2. The proofs of these properties are straightforward modifications of the proofs given in the appendix.

Definition 6.2.4 (Zero-coupon bond prices). *Let $\widetilde{\mathbb{P}}$ be a probability measure on the space Ω of all possible sequences of N coin tosses, and assume that every sequence $\omega_1 \dots \omega_N$ has strictly positive probability under $\widetilde{\mathbb{P}}$. Let $R_0, R_1, R_2, \dots, R_{N-1}$ be an interest rate process, with each R_n depending on only the first n coin tosses and satisfying (6.2.1). Define the discount process D_n, $n = 0, 1, \dots, N$, by (6.2.2). For $0 \le n \le m \le N$, the* price at time n of the zero-coupon bond maturing at time m *is defined to be*

$$B_{n,m} = \widetilde{\mathbb{E}}_n \left[\frac{D_m}{D_n} \right]. \tag{6.2.4}$$

According to (6.2.4), the price at time m of the zero-coupon bond maturing at time m is $B_{m,m} = 1$. We are choosing the face value 1 as a convenient normalization. If the face value of a bond is some other quantity C, then the value of this bond at time n before maturity m is $CB_{n,m}$. If $n = 0$, then (6.2.4) reduces to (6.2.3).

Remark 6.2.5. We may use the property of "taking out what is known" (Theorem 2.3.2(ii)) to rewrite (6.2.4) as

$$D_n B_{n,m} = \widetilde{\mathbb{E}}_n[D_m]. \tag{6.2.5}$$

From this formula, we see that the discounted bond price $D_n B_{n,m}$, $n = 0, 1, \dots, m$, is a martingale under $\widetilde{\mathbb{P}}$. In particular, if $0 \le k \le n \le m$, then (6.2.5) and "iterated conditioning" (Theorem 2.3.2(iii)) imply

$$\widetilde{\mathbb{E}}_k[D_n B_{n,m}] = \widetilde{\mathbb{E}}_k\left[\widetilde{\mathbb{E}}_n[D_m]\right] = \widetilde{\mathbb{E}}_k[D_m] = D_k B_{k,m},$$

which is the martingale property. Definition 6.2.4 was chosen so that discounted zero-coupon bond prices would be martingales.

In the discrete-time model of this chapter, investing in the money market is no different from investing in the bond maturing in one period (see Remark 6.2.9 below). Nonetheless, we treat the money market account as a separate instrument in order to write (6.2.6) below, which is a straightforward extension of (1.2.14). Consider an agent who can trade in the zero-coupon bonds of every maturity and in the money market. We wish to show that the wealth of such an agent, when discounted, constitutes a martingale. From this, we shall see that there is no arbitrage in the model in which zero-coupon bond prices are defined by (6.2.4). In particular, let $\Delta_{n,m}$ be the number of m-maturity zero-coupon bonds held by the agent between times n and $n+1$. Of course, we must have $n < m$ since the m-maturity bond does not exist after time m. Furthermore, $\Delta_{n,m}$ is allowed to depend on only the first n coin tosses. The agent begins with a nonrandom initial wealth X_0 and at each time n has wealth X_n. His wealth at time $n+1$ is thus given by

$$X_{n+1} = \Delta_{n,n+1} + \sum_{m=n+2}^{N} \Delta_{n,m} B_{n+1,m} + (1+R_n)\left(X_n - \sum_{m=n+1}^{N} \Delta_{n,m} B_{n,m}\right). \tag{6.2.6}$$

The first term on the right-hand side of (6.2.6) is the payoff of the zero-coupon bond maturing at time $n+1$, multiplied by the positions taken in this bond at time n and held to time $n+1$. The second term is the value of all the zero-coupon bonds maturing at time $n+2$ and later, multiplied by the position taken in these bonds at time n and held to $n+1$. The second factor in the third term is the cash position taken at time n, which is the difference between the total wealth X_n of the agent and the value of all the bonds held after rebalancing at time n. This is multiplied by 1 plus the interest rate that prevails between times n and $n+1$.

Theorem 6.2.6. *Regardless of how the portfolio random variables $\Delta_{n,m}$ are chosen (subject to the condition that $\Delta_{n,m}$ may depend only on the first n coin tosses), the discounted wealth process $D_n X_n$ is a martingale under $\widetilde{\mathbb{P}}$.*

PROOF: We use the fact that $\Delta_{n,m}$ depends on only the first n coin tosses and "taking out what is known" (Theorem 2.3.2(ii)), the fact that D_{n+1} also depends only on the first n coin tosses, and the martingale property of Remark 6.2.5 to compute

$$\begin{aligned}
\mathbb{E}_n[X_{n+1}] &= \Delta_{n,n+1} + \sum_{m=n+2}^{N} \Delta_{n,m}\mathbb{E}_n[B_{n+1,m}] \\
&\quad +(1+R_n)\left(X_n - \sum_{m=n+1}^{N} \Delta_{n,m}B_{n,m}\right) \\
&= \Delta_{n,n+1} + \sum_{m=n+2}^{N} \frac{\Delta_{n,m}}{D_{n+1}}\mathbb{E}_n[D_{n+1}B_{n+1,m}] \\
&\quad +\frac{D_n}{D_{n+1}}\left(X_n - \sum_{m=n+1}^{N} \Delta_{n,m}B_{n,m}\right) \\
&= \Delta_{n,n+1} + \sum_{m=n+2}^{N} \frac{\Delta_{n,m}}{D_{n+1}}D_nB_{n,m} + \frac{D_n}{D_{n+1}}X_n \\
&\quad -\frac{D_n}{D_{n+1}}\sum_{m=n+1}^{N} \Delta_{n,m}B_{n,m} \\
&= \Delta_{n,n+1} + \frac{D_n}{D_{n+1}}X_n - \frac{D_n}{D_{n+1}}\Delta_{n,n+1}B_{n,n+1}.
\end{aligned}$$

But $B_{n,n+1} = \widetilde{\mathbb{E}}_n\left[\frac{D_{n+1}}{D_n}\right] = \frac{D_{n+1}}{D_n}$ because $\frac{D_{n+1}}{D_n}$ depends on only the first n coin tosses. Substituting this into the equation above, we obtain

$$\widetilde{\mathbb{E}}_n[X_{n+1}] = \frac{D_n}{D_{n+1}}X_n.$$

Using again the fact that D_{n+1} depends only on the first n coin tosses, we may rewrite this as

$$\widetilde{\mathbb{E}}_n[D_{n+1}X_{n+1}] = D_nX_n,$$

which is the martingale property for the discounted wealth process. □

Remark 6.2.7. Because the discounted wealth process is a martingale under $\widetilde{\mathbb{P}}$, it has constant expectation:

$$\widetilde{\mathbb{E}}[D_nX_n] = X_0,\ n = 0, 1, \ldots, N. \tag{6.2.7}$$

If one could construct an arbitrage by trading in the zero-coupon bonds and the money market, then there would be a portfolio process that begins with $X_0 = 0$ and at some future time n results in $X_n \geq 0$, regardless of the outcome of the coin tossing, and further results in $X_n > 0$ for some of the outcomes. In such a situation, we would have $\widetilde{\mathbb{E}}[D_nX_n] > 0 = X_0$, a situation that is ruled out by (6.2.7). In other words, by using the risk-neutral pricing formula (6.2.4) to define zero-coupon bond prices, we have succeeded in building a model that is free of arbitrage.

Remark 6.2.8. The risk-neutral pricing formula says that for $0 \le n \le m \le N$, the price at time n of a derivative security paying V_m at time m (where V_m depends on only the first m coin tosses) is

$$V_n = \frac{1}{D_n}\widetilde{\mathbb{E}}_n[D_m V_m]. \tag{6.2.8}$$

Theorem 6.2.6 provides a partial justification for this. Namely, if it is possible to construct a portfolio that hedges a short position in the derivative security (i.e., that has value V_m at time m regardless of the outcome of the coin tossing), then the value of the derivative security at time n must be V_n given by (6.2.8). Theorem 6.2.6 does not guarantee that such a hedging portfolio can be constructed.

Remark 6.2.9. In the wealth equation (6.2.6), an agent is permitted to invest in the money market and zero-coupon bonds of all maturities. However, investing at time n in the zero-coupon bond with maturity $n+1$ is the same as investing in the money market. An investment of 1 in this bond at time n purchases $\frac{1}{B_{n,n+1}}$ bonds maturing at time $n+1$, and this investment pays off $\frac{1}{B_{n,n+1}}$ at time $n+1$. But

$$\frac{1}{B_{n,n+1}} = \frac{1}{\widetilde{\mathbb{E}}_n\left[\frac{D_{n+1}}{D_n}\right]} = \frac{1}{\widetilde{\mathbb{E}}_n\left[\frac{1}{1+R_n}\right]} = \frac{1}{\frac{1}{1+R_n}} = 1 + R_n.$$

This is the same payoff one would receive by investing 1 in the money market at time n. It is convenient to have a money market account and write (6.2.6) as we did, but it is actually unnecessary to include the money market account among the traded assets in this discrete-time model.

Let $0 \le m \le N$ be given. A *coupon-paying bond* can be modeled as a sequence of constant (i.e., nonrandom) quantities $C_0, C_1, \ldots, C_m$. For $0 \le n \le m-1$, the constant C_n is the coupon payment made at time n (which may be zero). The constant C_m is the payment made at time m, which includes principal as well as any coupon due at time m. In the case of the zero-coupon bond of Definition 6.2.4, $C_0 = C_1 = \cdots = C_{m-1} = 0$ and $C_m = 1$. In general, we may regard a coupon-paying bond as a sum of C_1 zero-coupon bonds maturing at time 1, C_2 zero-coupon bonds maturing at time 2, etc., up to C_m zero-coupon bonds maturing at time m. The price at time zero of the coupon-paying bond is thus

$$\sum_{k=0}^{m} C_k B_{0,k} = \widetilde{\mathbb{E}}\left[\sum_{k=0}^{m} D_k C_k\right].$$

At time n, before the payment C_n has been made but after the payments $C_0, \ldots, C_{n-1}$ have been made, the price of the coupon-paying bond is

$$\sum_{k=n}^{m} C_k B_{n,k} = \widetilde{\mathbb{E}}_n\left[\sum_{k=n}^{m} \frac{C_k D_k}{D_n}\right], \quad n = 0, 1, \ldots, m. \tag{6.2.9}$$

This generalizes formula (2.4.13) of Chapter 2 to the case of nonconstant interest rates but constant payment quantities $C_0, C_1, \ldots, C_m$.

6.3 Fixed-Income Derivatives

Suppose that in the binomial model for interest rates we have an asset whose price at time n we denote by S_n. This may be a stock but is more often a contract whose payoff depends on the interest rate. The price S_n is allowed to depend on the first n coin tosses. We have taken $\widetilde{\mathbb{P}}$ to be a risk-neutral measure, which means that, under $\widetilde{\mathbb{P}}$, the discounted asset price is a martingale:

$$D_n S_n = \widetilde{\mathbb{E}}_n \left[D_{n+1} S_{n+1} \right], \; n = 0, 1, \ldots, N-1. \tag{6.3.1}$$

Definition 6.3.1. *A* forward contract *is an agreement to pay a specified* delivery price K *at a* delivery date m, *where* $0 \leq m \leq N$, *for the asset whose price at time* m *is* S_m. *The* m-forward price *of this asset at time* n, *where* $0 \leq n \leq m$, *is the value of* K *that makes the forward contract have no-arbitrage price zero at time* n.

Theorem 6.3.2. *Consider an asset with price process* $S_0, S_1, \ldots, S_N$ *in the binomial interest rate model. Assume that zero-coupon bonds of all maturities can be traded. For* $0 \leq n \leq m \leq N$, *the* m*-forward price at time* n *of the asset is*

$$\text{For}_{n,m} = \frac{S_n}{B_{n,m}}. \tag{6.3.2}$$

PROOF: Suppose at time n an agent sells the forward contract with delivery date m and delivery price K. Suppose further that the value of K is chosen so that the forward contract has price zero at time n. Then, selling the forward contract generates no income. Having sold the forward contract at time n, suppose an agent immediately shorts $\frac{S_n}{B_{n,m}}$ zero-coupon bonds and uses the income S_n generated to buy one share of the asset. The agent then does no further trading until time m, at which time he owns one share of the asset, which he delivers according to the forward contract. In exchange, he receives K. After covering the short bond position, he is left with $K - \frac{S_n}{B_{n,m}}$. If this is positive, the agent has found an arbitrage. If it is negative, the agent could instead have taken the opposite positions, going long the forward, long the m-maturity bond, and short the asset, to again achieve an arbitrage. In order to preclude arbitrage, K must be given by (6.3.2). □

Remark 6.3.3. The proof of Theorem 6.3.2 constructs the hedge for a short position in a forward contract. This is called a *static hedge* because it calls for no trading between the time n when the hedge is set up and the time m when the forward contract expires. Whenever there is a hedge, static or not, the discounted value of the hedging portfolio is a martingale under the risk-neutral measure, and hence the risk-neutral pricing formula applies. In this case, that formula says the forward contract whose payoff is $S_m - K$ at time m must have time-n discounted price

$$\widetilde{\mathbb{E}}_n[D_m(S_m - K)] = \widetilde{\mathbb{E}}_n[D_m S_m] - K D_n \widetilde{\mathbb{E}}_n\left[\frac{D_m}{D_n}\right] = D_n(S_n - K B_{n,m}),$$

where we have used the martingale property (6.3.1) and the definition (6.2.4) of $B_{n,m}$. In order for the time-n price of the forward contract to be zero, we must have $S_n - KB_{n,m} = 0$, i.e., K is given by (6.3.2).

In addition to forward prices of assets, one can define forward interest rates. Here we show how to lock in forward one-period interest rates. Let $0 \le n \le m \le N-1$ be given. Suppose at time n an agent shorts an m-maturity zero-coupon bond and uses the income generated by this to purchase $\frac{B_{n,m}}{B_{n,m+1}}$ zero-coupon bonds maturing at time $m+1$. This portfolio is set up at zero cost at time n. At time m, it requires the agent to invest 1 to cover the short position in the m-maturity bond. At time $m+1$, the agent receives $\frac{B_{n,m}}{B_{n,m+1}}$ from the long position in the $m+1$-maturity bonds. Thus, between times m and $m+1$, it is as if the agent has invested at the interest rate

$$F_{n,m} = \frac{B_{n,m}}{B_{n,m+1}} - 1 = \frac{B_{n,m} - B_{n,m+1}}{B_{n,m+1}}. \tag{6.3.3}$$

Moreover, this one-period interest rate for investing between times m and $m+1$ was "locked in" by the portfolio set up at time n. Note that

$$F_{m,m} = \frac{B_{m,m}}{B_{m,m+1}} - 1 = \frac{1}{\frac{1}{1+R_m}} - 1 = R_m.$$

The one-period interest rate that can be locked in at time m for borrowing at time m is R_m.

Definition 6.3.4. *Let $0 \le n \le m \le N-1$ be given. The* forward one-period interest rate *at time n for investing at time m is defined by (6.3.3).*

Theorem 6.3.5. *Let $0 \le n \le m \le N-1$ be given. The no-arbitrage price at time n of a contract that pays R_m at time $m+1$ is $B_{n,m+1}F_{n,m} = B_{n,m} - B_{n,m+1}$.*

PROOF: Suppose at time n an agent sells a contract that promises to pay R_m at time $m+1$ and receives income $B_{n,m} - B_{n,m+1}$ for doing this. The agent then purchases one m-maturity bond and shorts an $m+1$-maturity bond. The total cost of setting up this portfolio, which is short one contract, long one bond, and short one bond, is zero. At time m, the agent receives income 1 for the m-maturity bond, and he invests this at the money market rate R_m. At time $m+1$, this investment yields $1+R_m$. The agent pays 1 to cover the short position in the $m+1$ maturity bond and uses the remaining R_m to pay off the contract. Thus, the agent has hedged the short position in the contract that promises to pay R_m at time $m+1$. The initial capital required to set up this hedge is $B_{n,m} - B_{n,m+1}$, which is thus the no-arbitrage price of the contract. □

According to Theorems 6.3.2 and 6.3.5, the forward price at time n of the contract that delivers R_m at time $m+1$ is $F_{n,m}$. (Replace m by $m+1$ in (6.3.2) and set $S_n = B_{n,m+1}F_{n,m}$.) This is another way to regard the concept of "locking in" an interest rate. Suppose at time n you short a forward contract on the interest rate (i.e., you agree to receive $F_{n,m}$ at time $m+1$ in exchange for a payment of R_m at that time). It costs nothing to enter this forward contract. Then you have locked in the interest rate $F_{n,m}$ at time m. Indeed, at time m you can invest 1 at the variable interest rate R_m. At time $m+1$, this investment yields $1+R_m$, but you have agreed to pay R_m in exchange for a payment of $F_{n,m}$, so the net amount you have at time $m+1$ after these settlements is $1+F_{n,m}$. You have effectively invested 1 between times m and $m+1$ at an interest rate of $F_{n,m}$.

Definition 6.3.6. *Let m be given with $1 \le m \le N$. An m-*period interest rate swap *is a contract that makes payments $S_1, \dots, S_m$ at times $1, \dots, m$, respectively, where*

$$S_n = K - R_{n-1}, \ n = 1, \dots, m.$$

The fixed rate *K is constant. The m-*period swap rate *SR_m is the value of K that makes the time-zero no-arbitrage price of the interest rate swap equal to zero.*

An agent with the long ("receive fixed") swap position receives a constant payment at each time n, a payment that one can regard as a fixed interest payment K on a principal amount of 1, and the agent makes a variable interest rate payment R_{n-1} on the same principal amount. If the agent already has a loan on which he is making fixed interest rate payments, the long swap position effectively converts this to a variable interest rate loan. A short swap position effectively converts a variable interest rate loan to a fixed interest rate loan.

Theorem 6.3.7. *The time-zero no-arbitrage price of the m-period interest rate swap in Definition 6.3.6 is*

$$\text{Swap}_m = \sum_{n=1}^{m} B_{0,n}\big(K - F_{0,n-1}\big) = K\sum_{n=1}^{m} B_{0,n} - \big(1 - B_{0,m}\big). \tag{6.3.4}$$

In particular, the m-period swap rate is

$$SR_m = \frac{\sum_{n=1}^{m} B_{0,n}F_{0,n-1}}{\sum_{n=1}^{m} B_{0,n}} = \frac{1 - B_{0,m}}{\sum_{n=1}^{m} B_{0,n}}. \tag{6.3.5}$$

PROOF: The time-zero no-arbitrage price of the payment K at time n is $KB_{0,n}$ and, according to Theorem 6.3.5, the time-zero no-arbitrage price of the payment R_{n-1} at time n is $B_{0,n}F_{0,n-1}$. Therefore, the time-zero price of S_n is $B_{0,n}\big(K - F_{0,n-1}\big)$. Summing, we obtain the middle term in (6.3.4). But

$$\sum_{n=1}^{m} B_{0,n}F_{0,n-1} = \sum_{n=1}^{m} \big(B_{0,n-1} - B_{0,n}\big) = 1 - B_{0,m}.$$

This observation gives us the right-hand side of (6.3.4). Setting the swap price equal to zero and solving for K, we obtain (6.3.5). □

Formula (6.3.4) for the no-arbitrage price of an m-period swap is of course also consistent with risk-neutral pricing. More specifically, we have from (6.3.4), (6.3.3), (6.2.4), and (6.2.2) that

$$\begin{aligned}
\text{Swap}_m &= \sum_{n=1}^{m} B_{0,n}(K - F_{0,n-1}) \\
&= \sum_{n=1}^{m} \big[KB_{0,n} - (B_{0,n-1} - B_{0,n})\big] \\
&= \sum_{n=1}^{m} \big[K\widetilde{\mathbb{E}}D_n - \big(\widetilde{\mathbb{E}}D_{n-1} - \widetilde{\mathbb{E}}D_n\big)\big] \\
&= \sum_{n=1}^{m} \big[K\widetilde{\mathbb{E}}D_n - \big(\widetilde{\mathbb{E}}[D_n(1 + R_{n-1})] - \widetilde{\mathbb{E}}D_n\big)\big] \\
&= \sum_{n=1}^{m} \big[K\widetilde{\mathbb{E}}D_n - \widetilde{\mathbb{E}}(D_nR_{n-1})\big] \\
&= \widetilde{\mathbb{E}}\sum_{n=1}^{m} D_n(K - R_{n-1}).
\end{aligned} \tag{6.3.6}$$

The last expression is the risk-neutral price of the swap.

Theorems 6.3.2, 6.3.5, and 6.3.7 determine prices of fixed income derivative securities by constructing hedging portfolios for short positions in the securities. Whenever such hedging portfolios can be constructed, risk-neutral pricing is justified, and (6.3.6) is an example of this. In the remainder of this section, we consider risk-neutral pricing of interest rate caps and floors, and although short position hedges for these instruments can usually be constructed, we do not work out the detailed assumptions that guarantee this is possible. Exercise 6.4 provides an example of a hedge construction.

Definition 6.3.8. *Let m be given, with $1 \le m \le N$. An m*-period interest rate cap *is a contract that makes payments $C_1, \ldots, C_m$ at times $1, \ldots, m$, respectively, where*

$$C_n = \big(R_{n-1} - K\big)^+, \; n = 1, \ldots, m.$$

An m-period interest rate floor *is a contract that makes payments* $F_1, \dots, F_m$ *at times* $1, \dots, m$, *respectively, where*

$$F_n = \left(K - R_{n-1}\right)^+, \ n = 1, \dots, m.$$

A contract that makes the payment C_n *at only one time* n *is called an* interest rate caplet, *and a contract that makes the payment* F_n *at only one time* n *is called an* interest rate floorlet. *The* risk-neutral price of an m-period interest rate cap *is*

$$\text{Cap}_m = \widetilde{\mathbb{E}} \sum_{n=1}^{m} D_n (R_{n-1} - K)^+, \tag{6.3.7}$$

and the risk-neutral price of an m-period interest rate floor *is*

$$\text{Floor}_m = \widetilde{\mathbb{E}} \sum_{n=1}^{m} D_n (K - R_{n-1})^+. \tag{6.3.8}$$

If one is paying the variable interest rate R_{n-1} at each time n on a loan of 1, then owning an interest rate cap effectively caps the interest rate at K. Whenever the interest owed is more than K, the cap pays the difference. Similarly, if one is receiving variable interest R_{n-1} at each time n on an investment of 1, then owning an interest rate floor effectively provides a guaranteed interest rate of at least K. Whenever the interest received is less than K, the floor pays the difference.

Note that

$$K - R_{n-1} + (R_{n-1} - K)^+ = (K - R_{n-1})^+.$$

In other words, at each time, the payoff of a portfolio holding a swap and a cap is the same as the payoff of a floor. It follows that

$$\text{Swap}_m + \text{Cap}_m = \text{Floor}_m. \tag{6.3.9}$$

In particular, if K is set equal to the m-period swap rate, then the cap and the floor have the same initial price.

Example 6.3.9. We return to the example of Figure 6.2.1 but now assign values to the interest rate in the tree as shown in Figure 6.3.1.

In Table 6.1 we show $D_1 = \frac{1}{1+R_0}$, $D_2 = \frac{1}{(1+R_0)(1+R_1)}$, and $D_3 = \frac{1}{(1+R_0)(1+R_1)(1+R_2)}$. The right-hand column of this table records the probability $\widetilde{\mathbb{P}}\{A_{\omega_1\omega_2}\}$; i.e., the first entry in this column is the probability of HH on the first two tosses.

We can then compute the time-zero bond prices:

$$B_{0,1} = \widetilde{\mathbb{E}} D_1 = 1,$$

$$B_{0,2} = \widetilde{\mathbb{E}} D_2 = \frac{6}{7} \cdot \frac{1}{3} + \frac{6}{7} \cdot \frac{1}{6} + \frac{5}{7} \cdot \frac{1}{4} + \frac{5}{7} \cdot \frac{1}{4} = \frac{11}{14},$$

$$B_{0,3} = \widetilde{\mathbb{E}} D_3 = \frac{3}{7} \cdot \frac{1}{3} + \frac{6}{7} \cdot \frac{1}{6} + \frac{4}{7} \cdot \frac{1}{4} + \frac{4}{7} \cdot \frac{1}{4} = \frac{4}{7}.$$

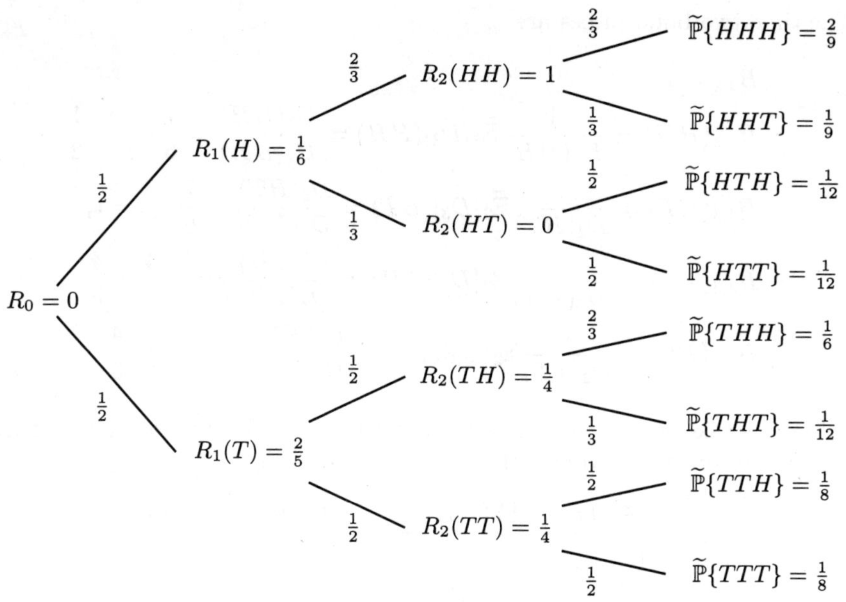

Fig. 6.3.1. A three-period interest rate model.

$\omega_1\omega_2$	$\frac{1}{1+R_0}$	$\frac{1}{1+R_1}$	$\frac{1}{1+R_2}$	D_1	D_2	D_3	$\widetilde{\mathbb{P}}$
HH	1	6/7	1/2	1	6/7	3/7	1/3
HT	1	6/7	1	1	6/7	6/7	1/6
TH	1	5/7	4/5	1	5/7	4/7	1/4
TT	1	5/7	4/5	1	5/7	4/7	1/4

Table 6.1.

The time-one bond prices are

$$\begin{aligned}
B_{1,1} &= 1, \\
B_{1,2}(H) &= \frac{1}{D_1(H)}\widetilde{\mathbb{E}}_1[D_2](H) = \frac{6}{7}\cdot\frac{2}{3} + \frac{6}{7}\cdot\frac{1}{3} = \frac{6}{7}, \\
B_{1,2}(T) &= \frac{1}{D_1(T)}\widetilde{\mathbb{E}}_1[D_2](T) = \frac{5}{7}\cdot\frac{1}{2} + \frac{5}{7}\cdot\frac{1}{2} = \frac{5}{7}, \\
B_{1,3}(H) &= \frac{1}{D_1(H)}\widetilde{\mathbb{E}}_1[D_3](H) = \frac{3}{7}\cdot\frac{2}{3} + \frac{6}{7}\cdot\frac{1}{3} = \frac{4}{7}, \\
B_{1,3}(T) &= \frac{1}{D_1(T)}\widetilde{\mathbb{E}}_1[D_3](T) = \frac{4}{7}\cdot\frac{1}{2} + \frac{4}{7}\cdot\frac{1}{2} = \frac{4}{7}.
\end{aligned}$$

The time-two bond prices are

$$B_{2,2} = 1,$$
$$B_{2,3}(HH) = \frac{1}{D_2(HH)}\widetilde{\mathbb{E}}_2[D_3](HH) = \frac{D_3(HH)}{D_2(HH)} = \frac{7}{6}\cdot\frac{3}{7} = \frac{1}{2},$$
$$B_{2,3}(HT) = \frac{1}{D_2(HT)}\widetilde{\mathbb{E}}_2[D_3](HT) = \frac{D_3(HT)}{D_2(HT)} = \frac{7}{6}\cdot\frac{6}{7} = 1,$$
$$B_{2,3}(TH) = \frac{1}{D_2(TH)}\widetilde{\mathbb{E}}_2[D_3](TH) = \frac{D_3(TH)}{D_2(TH)} = \frac{7}{5}\cdot\frac{4}{7} = \frac{4}{5},$$
$$B_{2,3}(TT) = \frac{1}{D_2(TT)}\widetilde{\mathbb{E}}_2[D_3](TT) = \frac{D_3(TT)}{D_2(TT)} = \frac{7}{5}\cdot\frac{4}{7} = \frac{4}{5}.$$

Let us take $K = \frac{1}{3}$ and price the corresponding three-period interest rate cap. We record the payoff of this cap in the following table.

$\omega_1\omega_2$	R_0	$\left(R_0 - \frac{1}{3}\right)^+$	R_1	$\left(R_1 - \frac{1}{3}\right)^+$	R_2	$\left(R_2 - \frac{1}{3}\right)^+$
HH	0	0	1/6	0	1	2/3
HT	0	0	1/6	0	0	0
TH	0	0	2/5	1/15	1/4	0
TT	0	0	2/5	1/15	1/4	0

The price at time zero of the time-one, time-two, and time-three caplets, respectively, are

$$\widetilde{\mathbb{E}}\left[D_1\left(R_0 - \frac{1}{3}\right)^+\right] = 0,$$
$$\widetilde{\mathbb{E}}\left[D_2\left(R_1 - \frac{1}{3}\right)^+\right] = \frac{5}{7}\cdot\frac{1}{15}\cdot\frac{1}{4} + \frac{5}{7}\cdot\frac{1}{15}\cdot\frac{1}{4} = \frac{1}{42},$$
$$\widetilde{\mathbb{E}}\left[D_3\left(R_2 - \frac{1}{3}\right)^+\right] = \frac{3}{7}\cdot\frac{2}{3}\cdot\frac{1}{3} = \frac{2}{21}, \tag{6.3.10}$$

and so $\text{Cap}_3 = 0 + \frac{1}{42} + \frac{2}{21} = \frac{5}{42}$.

6.4 Forward Measures

Let V_m be the payoff at time m of some contract (either a derivative security or some "primary" security such as a bond). The risk-neutral pricing formula for the price of this contract at times n prior to m involves the conditional expectation $\widetilde{\mathbb{E}}_n[D_m V_m]$. Indeed, the risk-neutral price of this security at time n is (see (6.2.8))

$$V_n = \frac{1}{D_n}\widetilde{\mathbb{E}}_n[D_m V_m], \quad n = 0, 1, \ldots, m. \tag{6.4.1}$$

To compute the conditional expectation appearing in this formula when D_m is random, one would need to know the *joint* conditional distribution of D_m and V_m under the risk-neutral measure $\widetilde{\mathbb{P}}$. This makes the pricing of fixed income derivatives difficult.

One way around this dilemma is to build interest rate models using *forward measures* rather than the risk-neutral measure. The idea is to use the term D_m on the right-hand side of (6.4.1) as a Radon-Nikodým derivative to change to a different measure, and provided we then compute expectations under this different measure, the term D_m no longer appears. To make this idea precise, we begin with the following definition.

Definition 6.4.1. *Let m be fixed, with $1 \leq m \leq N$. We define*

$$Z_{m,m} = \frac{D_m}{B_{0,m}} \tag{6.4.2}$$

*and use $Z_{m,m}$ to define the m-*forward measure *$\widetilde{\mathbb{P}}^m$ by the formula*

$$\widetilde{\mathbb{P}}^m(\omega) = Z_{m,m}(\omega)\widetilde{\mathbb{P}}(\omega) \text{ for all } \omega \in \Omega.$$

To check that $\widetilde{\mathbb{P}}^m$ is really a probability measure, we must verify that it assigns probability one to Ω. This is the case because $\widetilde{\mathbb{E}}Z_{m,m} = 1$. Indeed,

$$\widetilde{\mathbb{P}}^m(\Omega) = \sum_{\omega\in\Omega}\widetilde{\mathbb{P}}^m(\omega) = \sum_{\omega\in\Omega} Z_{m,m}(\omega)\widetilde{\mathbb{P}}(\omega) = \widetilde{\mathbb{E}}Z_{m,m} = \frac{1}{B_{0,m}}\widetilde{\mathbb{E}}D_m = 1,$$

where the last equality follows from the definition of zero-coupon bond prices (see (6.2.3)). Following the development in Section 3.2, we may define the *Radon-Nikodým derivative process*

$$Z_{n,m} = \widetilde{\mathbb{E}}_n Z_{m,m}, \quad n = 0, 1, \ldots, m. \tag{6.4.3}$$

If V_m is a random variable depending on only the first m coin tosses, then according to Lemma 3.2.5,

$$\widetilde{\mathbb{E}}^m V_m = \widetilde{\mathbb{E}}[Z_{m,m}V_m]. \tag{6.4.4}$$

More generally, if $0 \leq n \leq m$ and V_m depends on only the first m coin tosses, then according to Lemma 3.2.6,[1]

$$\widetilde{\mathbb{E}}_n^m[V_m] = \frac{1}{Z_{n,m}}\widetilde{\mathbb{E}}_n[Z_{m,m}V_m]. \tag{6.4.5}$$

[1] Although Lemmas 3.2.5 and 3.2.6 were developed for the transition from the actual probability measure $\mathbb{P}$ to the risk-neutral probability measure $\widetilde{\mathbb{P}}$, they apply to any transition from one probability measure to another. Here we use them to move from the risk-neutral probability measure $\widetilde{\mathbb{P}}$ to the forward measure $\widetilde{\mathbb{P}}^m$.

In the present context, with $Z_{m,m}$ defined by (6.4.2), equation (6.2.5) shows that equation (6.4.3) may be rewritten as

$$Z_{n,m} = \frac{D_n B_{n,m}}{B_{0,m}}, \; n = 0, \ldots, m. \tag{6.4.6}$$

Using (6.4.6) in (6.4.4) and (6.4.5), we obtain the following result.

Theorem 6.4.2. *Let m be fixed, with $1 \leq m \leq N$, and let $\widetilde{\mathbb{P}}^m$ denote the m-forward measure. If V_m is a random variable depending on only the first m coin tosses, then*

$$\widetilde{\mathbb{E}}^m[V_m] = \frac{1}{B_{0,m}} \widetilde{\mathbb{E}}[D_m V_m]. \tag{6.4.7}$$

More generally, if V_m depends on only the first m coin tosses, then

$$\widetilde{\mathbb{E}}^m_n[V_m] = \frac{1}{D_n B_{n,m}} \widetilde{\mathbb{E}}_n[D_m V_m], \; n = 0, 1, \ldots, m. \tag{6.4.8}$$

Computation of the left-hand side of (6.4.8) does not require us to know the correlation between V_m and D_m. Rather, we only need to know the conditional distribution of V_m under the m-forward measure $\widetilde{\mathbb{P}}^m$. Thus, the left-hand side of (6.4.8) is often easier to compute than the term $\widetilde{\mathbb{E}}_n[D_m V_m]$ appearing on the right-hand side. Note, however, that the m-forward measure is useful for pricing only those securities that pay off at time m, not at other times.

From (6.4.8) and (6.4.1), we have

$$\widetilde{\mathbb{E}}^m_n[V_m] = \frac{V_n}{B_{n,m}}, \; n = 0, 1, \ldots, m. \tag{6.4.9}$$

In other words, $\widetilde{\mathbb{E}}^m_n[V_m]$ is the price at time n of any derivative security or asset paying V_m at time m *denominated in units of the zero-coupon bond maturing at time m*. This is the *forward price* of the security or asset given in Theorem 6.3.2. The price of an asset denominated this way is a martingale under the forward measure $\widetilde{\mathbb{P}}^m$, as one can readily see by applying iterated conditioning to the left-hand side of (6.4.9). In conclusion, we see that *m-forward prices of (nondividend-paying) assets are martingales under the forward measure $\widetilde{\mathbb{P}}^m$.*

Example 6.4.3. Note from formula (6.4.2) that, like D_m, $Z_{m,m}$ depends on only the first $m-1$ coin tosses. We set $m = 3$ in Definition 6.4.1 and use the data in Example 6.3.9 so that $Z_{3,3}$ given by (6.4.2) is

$$Z_{3,3}(HH) = \frac{D_3(HH)}{B_{0,3}} = \frac{7}{4} \cdot \frac{3}{7} = \frac{3}{4}, \quad Z_{3,3}(HT) = \frac{D_3(HT)}{B_{0,3}} = \frac{7}{4} \cdot \frac{6}{7} = \frac{3}{2},$$

$$Z_{3,3}(TH) = \frac{D_3(TH)}{B_{0,3}} = \frac{7}{4} \cdot \frac{4}{7} = 1, \quad Z_{3,3}(TT) = \frac{D_3(TT)}{B_{0,3}} = \frac{7}{4} \cdot \frac{4}{7} = 1.$$

Note that $\widetilde{\mathbb{E}} Z_{3,3} = \frac{3}{4} \cdot \frac{1}{3} + \frac{3}{2} \cdot \frac{1}{6} + 1 \cdot \frac{1}{4} + 1 \cdot \frac{1}{4} = 1$, as it should. For each $\omega \in \Omega$, the values of $\widetilde{\mathbb{P}}(\omega)$, $Z_{3,3}(\omega)$, and $\widetilde{\mathbb{P}}^3(\omega) = Z_{3,3}(\omega)\widetilde{\mathbb{P}}(\omega)$ are given in the following table.

$\omega_1\omega_2\omega_3$	$\widetilde{\mathbb{P}}(\omega_1\omega_2\omega_3)$	$Z_{3,3}(\omega_1\omega_2\omega_3)$	$\widetilde{\mathbb{P}}^3(\omega_1\omega_2\omega_3)$
HHH	$\frac{2}{9}$	$\frac{3}{4}$	$\frac{1}{6}$
HHT	$\frac{1}{9}$	$\frac{3}{4}$	$\frac{1}{12}$
HTH	$\frac{1}{12}$	$\frac{3}{2}$	$\frac{1}{8}$
HTT	$\frac{1}{12}$	$\frac{3}{2}$	$\frac{1}{8}$
THH	$\frac{1}{6}$	1	$\frac{1}{6}$
THT	$\frac{1}{12}$	1	$\frac{1}{12}$
TTH	$\frac{1}{8}$	1	$\frac{1}{8}$
TTT	$\frac{1}{8}$	1	$\frac{1}{8}$

As in Example 6.2.1 and Figure 6.2.1, we can compute the $\widetilde{\mathbb{P}}^m$-conditional probabilities of getting H and T at each node in the tree representing the interest rate evolution. These conditional probabilities are shown in Figure 6.4.1.

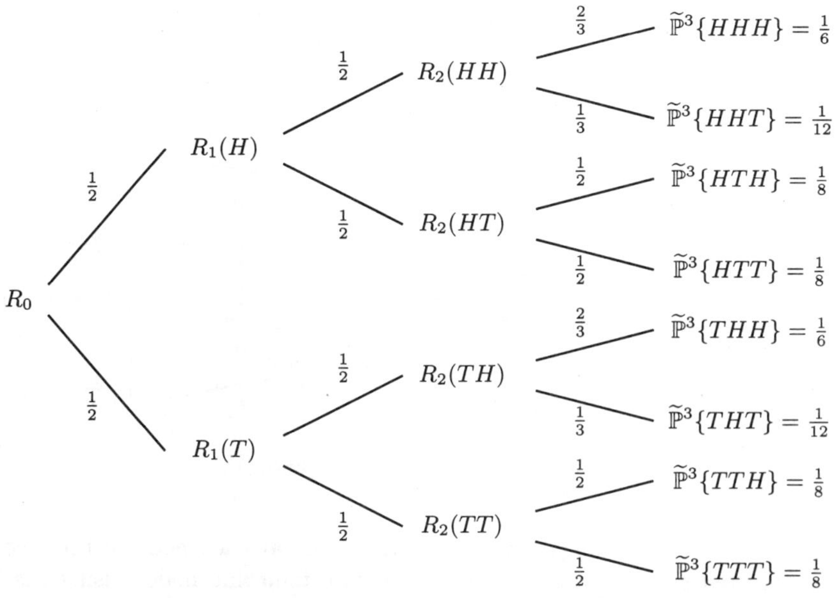

Fig. 6.4.1. Interest rate model and the $\widetilde{\mathbb{P}}^3$ transition probabilities.

We now compute the left-hand side of (6.4.8) with $V_3 = \left(R_2 - \frac{1}{3}\right)^+$, the payoff at time three of the caplet on the interest rate set at time two. In this case, V_3 depends on only the first two coin tosses and, in fact,

$$V_3(\omega_1\omega_2) = \begin{cases} \frac{2}{3}, \text{ if } \omega_1 = H, \omega_2 = H, \\ 0, \text{ otherwise.} \end{cases}$$

Since V_3 depends on only the first two tosses, we have $\widetilde{\mathbb{E}}_2^3[V_3] = V_3$. Using the probabilities shown in Figure 6.4.1, we compute

$$\widetilde{\mathbb{E}}_1^3[V_3](H) = \frac{1}{2}V_3(HH) + \frac{1}{2}V_3(HT) = \frac{1}{3},$$
$$\widetilde{\mathbb{E}}_1^3[V_3](T) = \frac{1}{2}V_3(TH) + \frac{1}{2}V_3(TT) = 0,$$
$$\widetilde{\mathbb{E}}^3[V_3] = \frac{1}{4}V_3(HH) + \frac{1}{4}V_3(HT) + \frac{1}{4}V_3(TH) + \frac{1}{4}V_3(TT) = \frac{1}{6}.$$

The process $\widetilde{\mathbb{E}}_n^3[V_3]$, $n = 0, 1, 2, 3$, is displayed in Figure 6.4.2. We note that

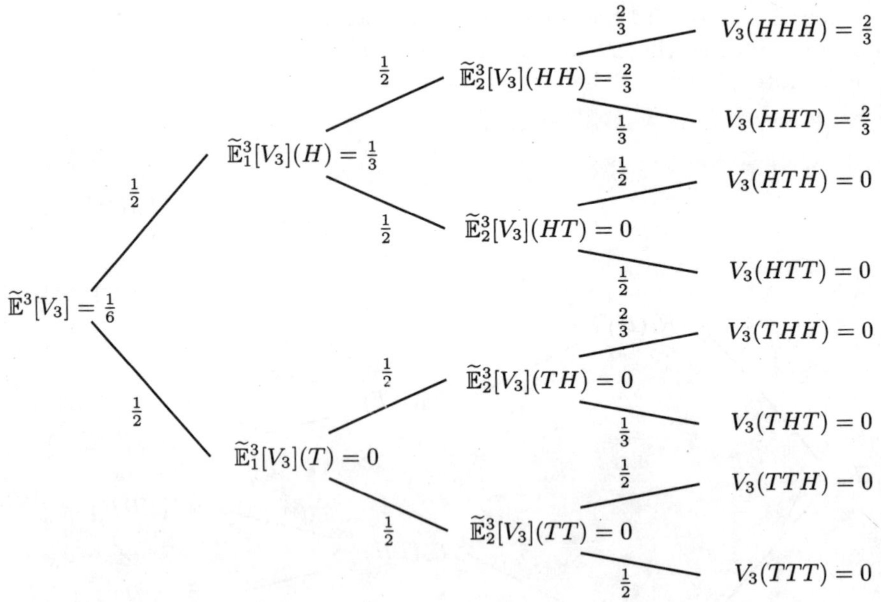

Fig. 6.4.2. The $\widetilde{\mathbb{P}}^3$ martingale $\widetilde{\mathbb{E}}_n^3[V_3]$.

this process is a martingale under $\widetilde{\mathbb{P}}^3$; the value at each node in the tree is the weighted average of the values at the two following nodes, using the $\widetilde{\mathbb{P}}^3$ transition probabilities shown on the links.

Finally, we use (6.4.9) to compute the time-zero price of the time-three caplet, which is

$$V_0 = B_{0,3}\widetilde{\mathbb{E}}^3[V_3] = \frac{4}{7} \cdot \frac{1}{6} = \frac{2}{21},$$

where we have used the bond price $B_{0,3} = \frac{4}{7}$ computed in Example 6.3.9. This agrees with (6.3.10). □

Example 6.4.4 (Ho-Lee model). In the Ho-Lee model [22], the interest rate at time n is

$$R_n(\omega_1 \dots \omega_n) = a_n + b_n \cdot \#H(\omega_1 \dots \omega_n),$$

where $a_0, a_1, \dots$ and $b_1, b_2, \dots$ are constants used to calibrate the model (i.e., make the prices generated by the model agree with market data). The risk-neutral probabilities are taken to be $\tilde{p} = \tilde{q} = \frac{1}{2}$.

In contrast to the kinds of numbers appearing in Examples 6.3.9 and 6.4.3, which were chosen to simplify the arithmetic, the Ho-Lee model can be used to generate numbers for practical applications. We repeat the computations of Examples 6.3.9 and 6.4.3 for the three-period Ho-Lee model shown in Figure 6.4.3 with $a_0 = 0.05$, $a_1 = 0.045$, $a_2 = 0.04$, and $b_1 = b_2 = 0.01$.

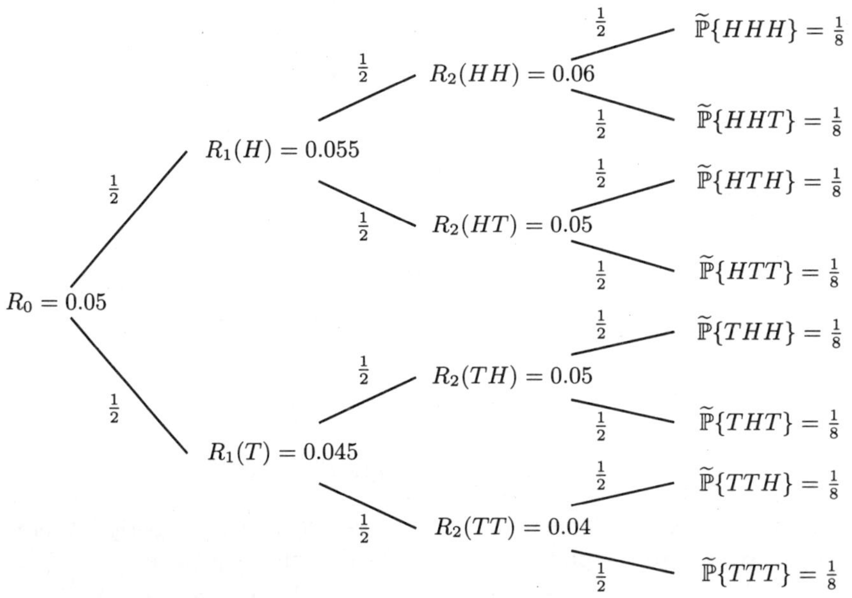

Fig. 6.4.3. A three-period Ho-Lee interest rate model.

In the following table, we show $D_1 = \frac{1}{1+R_0}$, $D_2 = \frac{1}{(1+R_0)(1+R_1)}$, and $D_3 = \frac{1}{(1+R_0)(1+R_1)(1+R_2)}$.

$\omega_1\omega_2$	$\frac{1}{1+R_0}$	$\frac{1}{1+R_1}$	$\frac{1}{1+R_2}$	D_1	D_2	D_3	$\widetilde{\mathbb{P}}$
HH	0.9524	0.9479	0.9434	0.9524	0.9027	0.8516	1/4
HT	0.9524	0.9479	0.9524	0.9524	0.9027	0.8597	1/4
TH	0.9524	0.9569	0.9524	0.9524	0.9114	0.8680	1/4
TT	0.9524	0.9569	0.9615	0.9524	0.9114	0.8763	1/4

We can then compute the time-zero bond prices:

$$B_{0,1} = \widetilde{\mathbb{E}}D_1 = 0.9524, \quad B_{0,2} = \widetilde{\mathbb{E}}D_2 = 0.9071, \quad B_{0,3} = \widetilde{\mathbb{E}}D_3 = 0.8639.$$

The time-one bond prices are $B_{1,1} = 1$ and

$$B_{1,2}(H) = \frac{1}{D_1(H)}\widetilde{\mathbb{E}}_1[D_2](H) = 0.9479,$$
$$B_{1,2}(T) = \frac{1}{D_1(T)}\widetilde{\mathbb{E}}_1[D_2](T) = 0.9569,$$
$$B_{1,3}(H) = \frac{1}{D_1(H)}\widetilde{\mathbb{E}}_1[D_3](H) = 0.8985,$$
$$B_{1,3}(T) = \frac{1}{D_1(T)}\widetilde{\mathbb{E}}_1[D_3](T) = 0.9158.$$

The time-two bond prices are $B_{2,2} = 1$ and

$$B_{2,3}(HH) = \frac{1}{D_2(HH)}\widetilde{\mathbb{E}}_2[D_3](HH) = 0.9434,$$
$$B_{2,3}(HT) = \frac{1}{D_2(HT)}\widetilde{\mathbb{E}}_2[D_3](HT) = 0.9524,$$
$$B_{2,3}(TH) = \frac{1}{D_2(TH)}\widetilde{\mathbb{E}}_2[D_3](TH) = 0.9524,$$
$$B_{2,3}(TT) = \frac{1}{D_2(TT)}\widetilde{\mathbb{E}}_2[D_3](TT) = 0.9615.$$

Let us take $K = 0.05$ and price the corresponding three-period interest rate cap. We record the payoff of this cap in the following table.

$\omega_1\omega_2$	R_0	$(R_0 - 0.05)^+$	R_1	$(R_1 - 0.05)^+$	R_2	$(R_2 - 0.05)^+$
HH	0.05	0	0.055	0.005	0.06	0.01
HT	0.05	0	0.055	0.005	0.05	0
TH	0.05	0	0.045	0	0.05	0
HT	0.05	0	0.045	0	0	0

The prices at time zero of the time-one, time-two, and time-three caplets, respectively, are

$$\widetilde{\mathbb{E}}\left[D_1\left(R_0 - 0.05\right)^+\right] = 0,$$
$$\widetilde{\mathbb{E}}\left[D_2\left(R_1 - 0.05\right)^+\right] = 0.002257,$$
$$\widetilde{\mathbb{E}}\left[D_3\left(R_2 - 0.05\right)^+\right] = 0.002129,$$

so the price of the cap consisting of these three caplets is $\text{Cap}_3 = 0.004386$.

We next compute the 3-forward measure $\widetilde{\mathbb{P}}^3$. The Radon-Nikodým derivative of this measure with respect to the risk-neutral measure $\widetilde{\mathbb{P}}$ is

$$Z_{3,3}(HH) = \frac{D_3(HH)}{B_{0,3}} = 0.9858, \quad Z_{3,3}(HT) = \frac{D_3(HT)}{B_{0,3}} = 0.9952,$$
$$Z_{3,3}(TH) = \frac{D_3(TH)}{B_{0,3}} = 1.0047, \quad Z_{3,3}(TT) = \frac{D_3(TT)}{B_{0,3}} = 1.0144.$$

Note that $\widetilde{\mathbb{E}}Z_{3,3} = \frac{1}{4}\big(Z_{3,3}(HH)+Z_{3,3}(HT)+Z_{3,3}(TH)+Z_{3,3}(TT)\big) = 1$, as it should. For each $\omega \in \Omega$, the values of $\widetilde{\mathbb{P}}(\omega)$, $Z_{3,3}(\omega)$, and $\widetilde{\mathbb{P}}^3(\omega) = Z_{3,3}(\omega)\widetilde{\mathbb{P}}(\omega)$ are given in the following table.

$\omega_1\omega_2\omega_3$	$\widetilde{\mathbb{P}}(\omega_1\omega_2\omega_3)$	$Z_{3,3}(\omega_1\omega_2\omega_3)$	$\widetilde{\mathbb{P}}^3(\omega_1\omega_2\omega_3)$
HHH	$\frac{1}{8}$	0.9858	0.1232
HHT	$\frac{1}{8}$	0.9858	0.1232
HTH	$\frac{1}{8}$	0.9952	0.1244
HTT	$\frac{1}{8}$	0.9952	0.1244
THH	$\frac{1}{8}$	1.0047	0.1256
THT	$\frac{1}{8}$	1.0047	0.1256
TTH	$\frac{1}{8}$	1.0144	0.1268
TTT	$\frac{1}{4}$	1.0144	0.1268

The $\widetilde{\mathbb{P}}^3$ transition probabilities are shown in Figure 6.4.4.

We now compute the left-hand side of (6.4.8) with $V_3 = \left(R_2 - 0.05\right)^+$, the payoff at time three of the caplet on the interest rate set at time two. In this case, V_3 depends on only the first two coin tosses and, in fact,

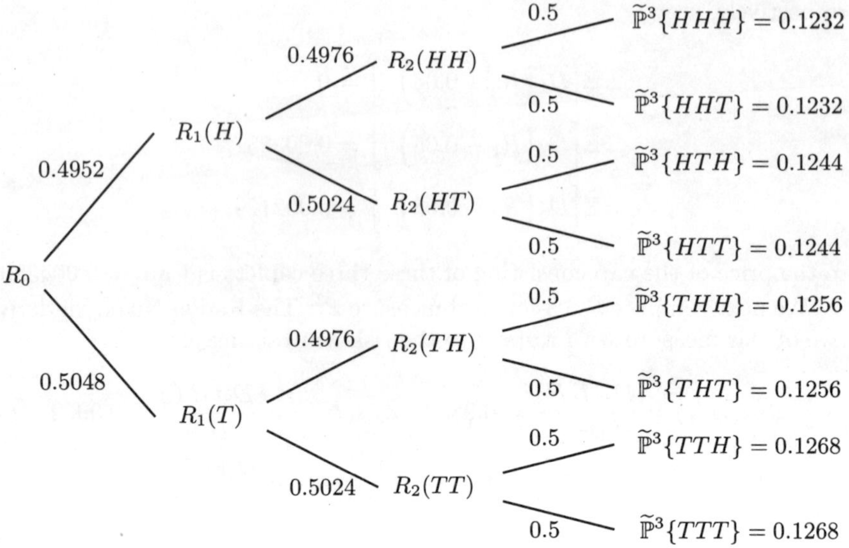

Fig. 6.4.4. The $\widetilde{\mathbb{P}}^3$ transition probabilities.

$$V_3(\omega_1\omega_2) = \begin{cases} 0.01, & \text{if } \omega_1 = H, \omega_2 = H, \\ 0, & \text{otherwise.} \end{cases}$$

Since V_3 depends on only the first two tosses, we have $\widetilde{\mathbb{E}}_2^3[V_3] = V_3$. Using the probabilities shown in Figure 6.4.4, we obtain the formulas shown in Figure 6.4.5. We note that this process is a martingale under $\widetilde{\mathbb{P}}^3$; the value at each node in the tree is the weighted average of the values at the two following nodes, using the $\widetilde{\mathbb{P}}^3$ transition probabilities shown on the links.

Finally, we use (6.4.9) to compute the time-zero price of the time-three caplet, which is $V_0 = B_{0,3}\widetilde{\mathbb{E}}^3[V_3] = (0.8639)(0.002464) = 0.002129.$ □

6.5 Futures

Like a forward contract, a *futures contract* is designed to lock in a price for purchase or sale of an asset before the time of the purchase or sale. It is designed to address two shortcomings of forward contracts. The first of these is that on any date prior to a specified delivery date, there can be demand for forward contracts with that delivery date. Hence, efficient markets would for each delivery date need a multitude of forward contracts with different initiation dates. By contrast, the underlying process for futures contracts is a *futures price* tied to a delivery date but not an initiation date, and all agents, regardless of the date they wish to enter a futures contract, trade on

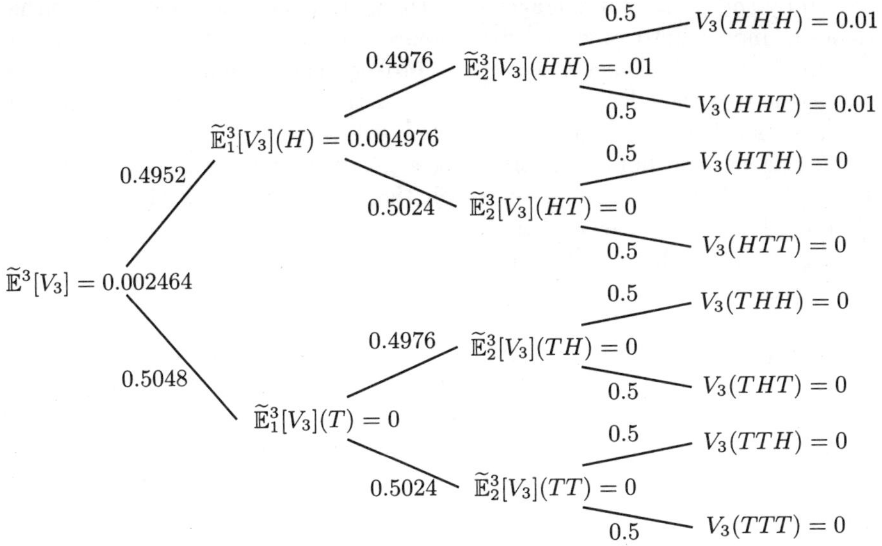

Fig. 6.4.5. The $\widetilde{\mathbb{P}}^3$ martingale $\widetilde{\mathbb{E}}_n^3[V_3]$.

the futures price for their desired delivery date. Secondly, although forward contracts have zero value at initiation, as the price of the underlying asset moves, the contract takes on a value that may be either positive or negative. In either case, one party may become concerned about default by the other party. Futures contracts, through *marking to market*, require agents to settle daily so that no agent ever has a large obligation to another.

Definition 6.5.1. *Consider an asset with price process $S_0, S_1, \ldots, S_N$ in the binomial interest rate model. For $0 \leq m \leq N$, the m-*futures price process $\text{Fut}_{n,m}$, *$n = 0, 1, \ldots, m$, is an adapted process with the following properties:*

(i) $\text{Fut}_{m,m} = S_m$;

(ii) For each n, $0 \leq n \leq m-1$, the risk-neutral value at time n of the contract that receives the payments $\text{Fut}_{k+1,m} - \text{Fut}_{k,m}$ at time $k+1$ for all $k = n, \ldots, m-1$, is zero; i.e.,

$$\frac{1}{D_n}\widetilde{\mathbb{E}}_n\left[\sum_{k=n}^{m-1} D_{k+1}(\text{Fut}_{k+1,m} - \text{Fut}_{k,m})\right] = 0. \tag{6.5.1}$$

An agent who at time n takes a long position in a futures contract with delivery at a later time m is agreeing to receive the payments $\text{Fut}_{k+1,m} - F_{k,m}$ at times $k+1$, for $k = n, \ldots, m-1$ and then to take delivery of the asset at its market price S_m at time m. Any of the payments $\text{Fut}_{k+1,m} - \text{Fut}_{k,m}$ can be negative, in which case the agent in the long position must pay money rather than receive it. The vehicle for these transfers of money is the margin account set up by the broker for the agent. At the end of each day, the broker either deposits money into or withdraws money from the margin account. If the account balance becomes too low, the broker issues a margin call, requiring the agent to either deposit more money in the account or close the futures position (see below).

An agent who holds a short futures position makes payments equal to $(\mathrm{Fut}_{k+1,m} - \mathrm{Fut}_{k,m})$ and agrees to sell the asset at its market price S_m at time m. This agent likewise has a margin account.

The agent holding the long position between times n and m receives total payments

$$\sum_{k=n}^{m-1} (\mathrm{Fut}_{k+1,m} - \mathrm{Fut}_{k,m}) = \mathrm{Fut}_{m,m} - \mathrm{Fut}_{n,m} = S_m - \mathrm{Fut}_{n,m}\,.$$

After purchasing the asset at the market price S_m, the agent has a balance of

$$S_m - \mathrm{Fut}_{n,m} - S_m = -\mathrm{Fut}_{n,m}$$

from the futures trading and the asset purchase. In addition, the agent owns the asset. In effect, if one ignores the time value of money, the agent has paid $F_{n,m}$ and acquired the asset. In this sense, a price was locked in at time n for purchase at the later time m.

Condition (6.5.1) is designed so that at the time of initiation, the value of the futures contract is zero. Thus, it costs nothing to enter a long (or a short) futures position, apart from the requirement that a margin account be set up. Since (6.5.1) is required to hold for all n between 0 and $m-1$, the value of the futures contract is zero at all times. In particular, it costs nothing for an agent who holds a long position to "sell" his contract (i.e., to *close* the position). At that time, all payments (positive and negative) to the agent cease and the agent no longer is obligated to purchase the asset at the market price at time m. Similarly, an agent with a short futures position can close the position at any time at no cost.

Theorem 6.5.2. *Let m be given with $0 \le m \le N$. Then*

$$\mathrm{Fut}_{n,m} = \widetilde{\mathbb{E}}_n[S_m], \; n = 0, 1, \ldots, m \tag{6.5.2}$$

is the unique process satisfying the conditions of Definition 6.5.1.

PROOF: We first show that the right-hand side of (6.5.2) satisfies the conditions of Definition 6.5.1. It is clear that $\widetilde{\mathbb{E}}_m[S_m] = S_m$, which shows that the right-hand side of (6.5.2) satisfies Definition 6.5.1(i). To verify Definition 6.5.1(ii), it suffices to show that each term in the sum in (6.5.1) has $\widetilde{\mathbb{E}}_n$ conditional expectation zero; i.e.,

$$\widetilde{\mathbb{E}}_n\big[D_{k+1}\big(\widetilde{\mathbb{E}}_{k+1}[S_m] - \widetilde{\mathbb{E}}_k[S_m]\big)\big] = 0, \; k = n, \ldots, m-1. \tag{6.5.3}$$

Because D_{k+1} depends on only the first k coin tosses, for $k = n, \ldots, m-1$, iterated conditioning implies

$$\begin{aligned}
\widetilde{\mathbb{E}}_n\Big[D_{k+1}\Big(\widetilde{\mathbb{E}}_{k+1}[S_m] - \widetilde{\mathbb{E}}_k[S_m]\Big)\Big] &= \widetilde{\mathbb{E}}_n\Big[\widetilde{\mathbb{E}}_k\Big[D_{k+1}\Big(\widetilde{\mathbb{E}}_{k+1}[S_m] - \widetilde{\mathbb{E}}_k[S_m]\Big)\Big]\Big] \\
&= \widetilde{\mathbb{E}}_n\Big[D_{k+1}\Big(\widetilde{\mathbb{E}}_k\big[\widetilde{\mathbb{E}}_{k+1}[S_m]\big] - \widetilde{\mathbb{E}}_k[S_m]\Big)\Big] \\
&= \widetilde{\mathbb{E}}_n\Big[\Big(D_{k+1}\big(\widetilde{\mathbb{E}}_k[S_m] - \widetilde{\mathbb{E}}_k[S_m]\big)\Big)\Big] \\
&= 0,
\end{aligned}$$

and (6.5.3) is established.

We next show that the right-hand side of (6.5.2) is the only process that satisfies the conditions of Definition 6.5.1. From (6.5.1), we have

$$\sum_{k=n}^{m-1} \widetilde{\mathbb{E}}_n\big[D_{k+1}(\mathrm{Fut}_{k+1,m} - \mathrm{Fut}_{k,m})\big] = 0 \text{ for } n = 0, 1, \ldots, m-1. \tag{6.5.4}$$

For $n = 0, 1, \ldots, m-2$, we may replace n by $n+1$ in this equation and subtract to obtain

$$\begin{aligned}
0 = \sum_{k=n}^{m-1} &\widetilde{\mathbb{E}}_n\big[D_{k+1}(\mathrm{Fut}_{k+1,m} - \mathrm{Fut}_{k,m})\big] \\
&- \sum_{k=n+1}^{m-1} \widetilde{\mathbb{E}}_{n+1}\big[D_{k+1}(\mathrm{Fut}_{k+1,m} - \mathrm{Fut}_{k,m})\big].
\end{aligned} \tag{6.5.5}$$

If we now take conditional expectations $\widetilde{\mathbb{E}}_n$ in (6.5.5) and use iterated conditioning, we see that

$$\begin{aligned}
0 = \sum_{k=n}^{m-1} &\widetilde{\mathbb{E}}_n\big[D_{k+1}(\mathrm{Fut}_{k+1,m} - \mathrm{Fut}_{k,m})\big] \\
&- \sum_{k=n+1}^{m-1} \widetilde{\mathbb{E}}_n\big[D_{k+1}(\mathrm{Fut}_{k+1,m} - \mathrm{Fut}_{k,m})\big] \\
= \widetilde{\mathbb{E}}_n&\big[D_{n+1}(\mathrm{Fut}_{n+1,m} - \mathrm{Fut}_{n,m})\big].
\end{aligned} \tag{6.5.6}$$

Setting $n = m-1$ in (6.5.4), we see that (6.5.6) also holds for $n = m-1$. Both $\mathrm{Fut}_{n,m}$ and D_{n+1} depend on only the first n coin tosses, and hence (6.5.6) reduces to

$$D_{n+1}\big(\widetilde{\mathbb{E}}_n[\mathrm{Fut}_{n+1,m}] - \mathrm{Fut}_{n,m}\big) = 0,$$

which yields

$$\widetilde{\mathbb{E}}_n[\mathrm{Fut}_{n+1,m}] = \mathrm{Fut}_{n,m}, \; n = 0, 1, \ldots, m-1.$$

This is the martingale property under $\widetilde{\mathbb{P}}$ for $\mathrm{Fut}_{n,m}$, $n = 0, 1, \ldots, m$. But $\mathrm{Fut}_{m,m} = S_m$, and the martingale property implies (6.5.2). □

Corollary 6.5.3. *Let m be given with $0 \leq m \leq N$. Then $\mathrm{For}_{0,m} = \mathrm{Fut}_{0,m}$ if and only if D_m and S_m are uncorrelated under $\widetilde{\mathbb{P}}$. In particular, this is the case if the interest rate is not random.*

PROOF: From Theorem 6.3.2 we have

$$\mathrm{For}_{0,m} = \frac{S_0}{B_{0,m}} = \frac{S_0}{\widetilde{\mathbb{E}}D_m} = \frac{\widetilde{\mathbb{E}}[D_m S_m]}{\widetilde{\mathbb{E}}D_m}$$

and

$$\mathrm{Fut}_{0,m} = \widetilde{\mathbb{E}}S_m.$$

These two formulas agree if and only if $\widetilde{\mathbb{E}}[D_m S_m] = \widetilde{\mathbb{E}}D_m \cdot \widetilde{\mathbb{E}}S_m$, which is uncorrelatedness of D_m and S_m. □

Remark 6.5.4. We note in connection with Corollary 6.5.3 that if D_m and S_m are negatively correlated, so that $\widetilde{\mathbb{E}}[D_m S_m] < \widetilde{\mathbb{E}}D_m \cdot \widetilde{\mathbb{E}}S_m$, then $\mathrm{For}_{0,m} < \mathrm{Fut}_{0,m}$. This is the case when an increase in the asset price tends to be accompanied by an increase in the interest rate (so that there is a decrease in the discount factor). The long futures position benefits from this more than the long forward position since the long futures position receives an immediate payment to invest at the higher interest rate, whereas the forward contract has no settlement until the delivery date m. Because the long futures position is more favorable, the initial futures price is higher (i.e., the person using futures to lock in a purchase price locks in a higher purchase price than a person using a long forward position).

Example 6.5.5 (Black-Derman-Toy model). In the Black-Derman-Toy (BDT) model [4], the interest rate at time n is

$$R_n(\omega_1 \ldots \omega_n) = a_n b_n^{\#H(\omega_1 \ldots \omega_n)},$$

where the constants $a_0, a_1, \ldots$ and $b_1, b_2 \ldots$ are used to calibrate the model. The risk-neutral probabilities are taken to be $\tilde{p} = \tilde{q} = \frac{1}{2}$. With $a_n = \frac{0.05}{1.2^n}$ and $b_n = 1.44$, the three-period BDT model is shown in Figure 6.5.1. In this model, we have the following zero-coupon bond prices:

$$B_{0,2} = 0.9064, \;\; B_{0,3} = 0.8620,$$
$$B_{1,2}(H) = 0.9434, \;\; B_{1,2}(T) = 0.9600,$$
$$B_{1,3}(H) = 0.8893, \;\; B_{1,3}(T) = 0.9210.$$

From these, we can compute the forward interest rates $F_{n,2} = \frac{B_{n,2} - B_{n,3}}{B_{n,3}}$, obtaining

$$F_{0,2} = 0.05147, \quad F_{1,2}(H) = 0.06089, \quad F_{1,2}(T) = 0.04231.$$

These are the forward prices for a contract that pays R_2 at time three. (The second subscript 2 on these forward interest rates denotes the time the rate

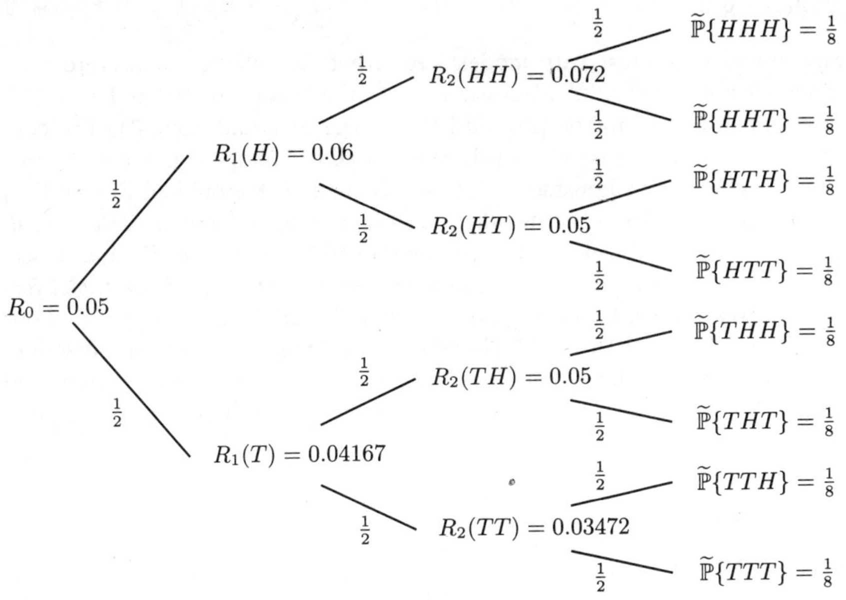

Fig. 6.5.1. A three-period Black-Derman-Toy interest rate model.

is set, not the time of payment.) The futures prices for the contract that pays R_2 at time three are given by $\mathrm{Fut}_{n,3} = \widetilde{\mathbb{E}}_n[R_2]$, and these turn out to be

$$\mathrm{Fut}_{0,3} = 0.05168, \quad \mathrm{Fut}_{1,3}(H) = 0.06100, \quad \mathrm{Fut}_{1,3}(T) = 0.04236.$$

The interest rate futures are slightly higher than the forward interest rates for the reason discussed in Remark 6.5.4.

6.6 Summary

Within the simple context of a binomial model, this chapter introduces several ideas that arise when interest rates are random. The prices of zero-coupon bonds are defined via the risk-neutral pricing formula (Definition 6.2.4), and it is shown that this method of pricing rules out arbitrage (Remark 6.2.7). Forward contracts, interest rate caps and floors, and futures contracts are then introduced. The last of these, appearing in Section 6.5, is related to forward contracts in Corollary 6.5.3.

Section 6.4 presents the concept of *forward measure*, an idea used in continuous-time models to simplify the pricing of interest rate caps and floors. We illustrate this idea in Example 6.4.3, but the full power of this approach is realized only in continuous-time models.

6.7 Notes

Many practical interest rate models are either formulated in discrete time or have discrete-time implementations (e.g., the models of Ho and Lee [22], Heath, Jarrow, and Morton [21], and Black, Derman, and Toy [4]). The concept of forward measures, which will be revisited in Volume II in a continuous-time setting, is due to Jamshidian [24] and Geman, El Karoui and Rochet [15]. These measures underlie the so-called *market models*, or *forward LIBOR models*, of interest rates developed by Sandmann and Sondermann [38] and Brace, Gatarek, and Musiela [6], models that are consistent with the *Black caplet formula.* The distinction between forward contracts and futures was pointed out by Margrabe [31] and Black [3]. No-arbitrage pricing of futures in a discrete-time model was developed by Cox, Ingersoll, and Ross [9] and Jarrow and Oldfield [25]. Exercise 6.7 is taken from Duffie [14], a highly recommended text.

6.8 Exercises

Exercise 6.1. Prove parts (i), (ii), (iii) and (v) of Theorem 2.3.2 when conditional expectation is defined by Definition 6.2.2. (Part (iv) is not true in the form stated in Theorem 2.3.2 when the coin tosses are not independent.)

Exercise 6.2. Verify that the discounted value of the static hedging portfolio constructed in the proof of Theorem 6.3.2 is a martingale under $\widetilde{\mathbb{P}}$.

Exercise 6.3. Let $0 \leq n \leq m \leq N-1$ be given. According to the risk-neutral pricing formula, the contract that pays R_m at time $m+1$ has time-n price $\frac{1}{D_n}\widetilde{\mathbb{E}}_n[D_{m+1}R_m]$. Use the properties of conditional expectations to show that this gives the same result as Theorem 6.3.5, i.e.,

$$\frac{1}{D_n}\widetilde{\mathbb{E}}_n[D_{m+1}R_m] = B_{n,m} - B_{n,m+1}.$$

Exercise 6.4. Using the data in Example 6.3.9, this exercise constructs a hedge for a short position in the caplet paying $(R_2 - \frac{1}{3})^+$ at time three. We observe from the second table in Example 6.3.9 that the payoff at time three of this caplet is

$$V_3(HH) = \frac{2}{3}, \quad V_3(HT) = V_3(TH) = V_3(TT) = 0.$$

Since this payoff depends on only the first two coin tosses, the price of the caplet at time two can be determined by discounting:

$$V_2(HH) = \frac{1}{1+R_2(HH)}V_3(HH) = \frac{1}{3}, \quad V_2(HT) = V_2(TH) = V_2(TT) = 0.$$

Indeed, if one is hedging a short position in the caplet and has a portfolio valued at $\frac{1}{3}$ at time two in the event $\omega_1 = H, \omega_2 = H$, then one can simply invest this $\frac{1}{3}$ in the money market in order to have the $\frac{2}{3}$ required to pay off the caplet at time three.

In Example 6.3.9, the time-zero price of the caplet is determined to be $\frac{2}{21}$ (see (6.3.10)).

(i) Determine $V_1(H)$ and $V_1(T)$, the price at time one of the caplet in the events $\omega_1 = H$ and $\omega_1 = T$, respectively.
(ii) Show how to begin with $\frac{2}{21}$ at time zero and invest in the money market and the maturity two bond in order to have a portfolio value X_1 at time one that agrees with V_1, regardless of the outcome of the first coin toss. Why do we invest in the maturity two bond rather than the maturity three bond to do this?
(iii) Show how to take the portfolio value X_1 at time one and invest in the money market and the maturity three bond in order to have a portfolio value X_2 at time two that agrees with V_2, regardless of the outcome of the first two coin tosses. Why do we invest in the maturity three bond rather than the maturity two bond to do this?

Exercise 6.5. Let m be given with $0 \le m \le N-1$, and consider the forward one-period interest rate

$$F_{n,m} = \frac{B_{n,m} - B_{n,m+1}}{B_{n,m+1}}, \quad n = 0, 1, \ldots, m.$$

(i) Use (6.4.8) and (6.2.5) to show that $F_{n,m}$, $n = 0, 1, \ldots, m$, is a martingale under the $(m+1)$-forward measure $\widetilde{\mathbb{P}}^{m+1}$.
(ii) Compute $F_{0,2}$, $F_{1,2}(H)$, and $F_{1,2}(T)$ in Example 6.4.4 and verify the martingale property

$$\widetilde{\mathbb{E}}^3[F_{1,2}] = F_{0,2}.$$

Exercise 6.6. Let S_m be the price at time m of an asset in a binomial interest rate model. For $n = 0, 1, \ldots, m$, the forward price is $\text{For}_{n,m} = \frac{S_n}{B_{n,m}}$ and the futures price is $\text{Fut}_{n,m} = \widetilde{\mathbb{E}}_n[S_n]$.

(i) Suppose that at each time n an agent takes a long forward position and sells this contract at time $n+1$. Show that this generates cash flow $S_{n+1} - \frac{S_n B_{n+1,m}}{B_{n,m}}$ at time $n+1$.
(ii) Show that if the interest rate is a constant r and at each time n an agent takes a long position of $(1+r)^{m-n-1}$ forward contracts, selling these contracts at time $n+1$, then the resulting cash flow is the same as the difference in the futures price $\text{Fut}_{n+1,m} - \text{Fut}_{n,m}$.

Exercise 6.7. Consider a binomial interest rate model in which the interest rate at time n depends on only the number of heads in the first n coin tosses. In other words, for each n there is a function $r_n(k)$ such that

$$R_n(\omega_1 \ldots \omega_n) = r_n\big(\#H(\omega_1 \ldots \omega_n)\big).$$

Assume the risk-neutral probabilities are $\tilde{p} = \tilde{q} = \frac{1}{2}$. The Ho-Lee model (Example 6.4.4) and the Black-Derman-Toy model (Example 6.5.5) satisfy these conditions.

Consider a derivative security that pays 1 at time n if and only if there are k heads in the first n tosses; i.e., the payoff is $V_n(k) = \mathbb{I}_{\{\#H(\omega_1 \ldots \omega_n)=k\}}$. Define $\psi_0(0) = 1$ and, for $n = 1, 2, \ldots$, define

$$\psi_n(k) = \widetilde{\mathbb{E}}[D_n V_n(k)], \quad k = 0, 1, \ldots, n,$$

to be the price of this security at time zero. Show that the functions $\psi_n(k)$ can be computed by the recursion

$$\begin{aligned}
\psi_{n+1}(0) &= \frac{\psi_n(0)}{2(1 + r_n(0))}, \\
\psi_{n+1}(k) &= \frac{\psi_n(k-1)}{2(1 + r_n(k-1))} + \frac{\psi_n(k)}{2(1 + r_n(k))}, \quad k = 1, \ldots, n, \\
\psi_{n+1}(n+1) &= \frac{\psi_n(n)}{2(1 + r_n(n))}.
\end{aligned}$$

A

Proof of Fundamental Properties of Conditional Expectations

This appendix provides the proof of Theorem 2.3.2 of Chapter 2, which is restated below.

Theorem 2.3.2 (Fundamental properties of conditional expectations). *Let N be a positive integer, and let X and Y be random variables depending on the first N coin tosses. Let $0 \leq n \leq N$ be given. The following properties hold.*

(i) ***Linearity of conditional expectations****. For all constants c_1 and c_2, we have*

$$\mathbb{E}_n[c_1X + c_2Y] = c_1\mathbb{E}_n[X] + c_2\mathbb{E}_n[Y].$$

(ii) ***Taking out what is known****. If X actually depends only on the first n coin tosses, then*

$$\mathbb{E}_n[XY] = X \cdot \mathbb{E}_n[Y].$$

(iii) ***Iterated conditioning****. If $0 \leq n \leq m \leq N$, then*

$$\mathbb{E}_n\Big[\mathbb{E}_m[X]\Big] = \mathbb{E}_n[X].$$

In particular, $\mathbb{E}\Big[\mathbb{E}_m[X]\Big] = \mathbb{E}X$.

(iv) ***Independence****. If X depends only on tosses $n+1$ through N, then*

$$\mathbb{E}_n[X] = \mathbb{E}X.$$

(v) ***Conditional Jensen's inequality****. If $\varphi(x)$ is a convex function of the dummy variable x, then*

$$\mathbb{E}_n[\varphi(X)] \geq \varphi(\mathbb{E}_n[X]).$$

PROOF: We start by recalling the definition of conditional expectation:

$$\begin{aligned}&\mathbb{E}_n[X](\omega_1 \dots \omega_n)\\ &= \sum_{\omega_{n+1}\dots\omega_N} p^{\#H(\omega_{n+1}\dots\omega_N)} q^{\#T(\omega_{n+1}\dots\omega_N)} X(\omega_1 \dots \omega_n\omega_{n+1} \dots \omega_N).\end{aligned}$$

Proof of (i):

$$
\begin{aligned}
&\mathbb{E}_n[c_1X + c_2Y](\omega_1 \ldots \omega_n)\\
&= \sum_{\omega_{n+1}\ldots\omega_N} p^{\#H(\omega_{n+1}\ldots\omega_N)} q^{\#T(\omega_{n+1}\ldots\omega_N)} [c_1X(\omega_1 \ldots \omega_N) + c_2Y(\omega_1 \ldots \omega_N)]\\
&= c_1 \sum_{\omega_{n+1}\ldots\omega_N} p^{\#H(\omega_{n+1}\ldots\omega_N)} q^{\#T(\omega_{n+1}\ldots\omega_N)} X(\omega_1 \ldots \omega_N)\\
&\quad + c_2 \sum_{\omega_{n+1}\ldots\omega_N} p^{\#H(\omega_{n+1}\ldots\omega_N)} q^{\#T(\omega_{n+1}\ldots\omega_N)} Y(\omega_1 \ldots \omega_N)\\
&= c_1\mathbb{E}_n[X](\omega_1 \ldots \omega_n) + c_2\mathbb{E}_n[Y](\omega_1 \ldots \omega_n).
\end{aligned}
$$

Proof of (ii):

$$
\begin{aligned}
&\mathbb{E}_n[XY](\omega_1 \ldots \omega_n)\\
&= \sum_{\omega_{n+1}\ldots\omega_N} p^{\#H(\omega_{n+1}\ldots\omega_N)} q^{\#T(\omega_{n+1}\ldots\omega_N)} X(\omega_1 \ldots \omega_n) Y(\omega_1 \ldots \omega_n\omega_{n+1} \ldots \omega_N)\\
&= X(\omega_1 \ldots \omega_n) \sum_{\omega_{n+1}\ldots\omega_N} p^{\#H(\omega_{n+1}\ldots\omega_N)} q^{\#T(\omega_{n+1}\ldots\omega_N)} Y(\omega_1 \ldots \omega_n\omega_{n+1} \ldots \omega_N)\\
&= X(\omega_1 \ldots \omega_n)\mathbb{E}_n[Y](\omega_1 \ldots \omega_n).
\end{aligned}
$$

Proof of (iii): Denote $Z = E_m[X]$. Then Z actually depends on $\omega_1\omega_2 \ldots \omega_m$ only and

$$
\begin{aligned}
&\mathbb{E}_n[\mathbb{E}_m[X]](\omega_1 \ldots \omega_n) = \mathbb{E}_n[Z](\omega_1 \ldots \omega_n)\\
&= \sum_{\omega_{n+1}\ldots\omega_N} p^{\#H(\omega_{n+1}\ldots\omega_N)} q^{\#T(\omega_{n+1}\ldots\omega_N)} Z(\omega_1 \ldots \omega_n\omega_{n+1} \ldots \omega_m)\\
&= \sum_{\omega_{n+1}\ldots\omega_m} p^{\#H(\omega_{n+1}\ldots\omega_m)} q^{\#T(\omega_{n+1}\ldots\omega_m)} Z(\omega_1 \ldots \omega_m)\\
&\qquad \cdot \sum_{\omega_{m+1}\ldots\omega_N} p^{\#H(\omega_{m+1}\ldots\omega_N)} q^{\#T(\omega_{m+1}\ldots\omega_N)}\\
&= \sum_{\omega_{n+1}\ldots\omega_m} p^{\#H(\omega_{n+1}\ldots\omega_m)} q^{\#T(\omega_{n+1}\ldots\omega_m)} Z(\omega_1 \ldots \omega_m)\\
&= \sum_{\omega_{n+1}\ldots\omega_m} p^{\#H(\omega_{n+1}\ldots\omega_m)} q^{\#T(\omega_{n+1}\ldots\omega_m)}\\
&\qquad \cdot \sum_{\omega_{m+1}\ldots\omega_N} p^{\#H(\omega_{m+1}\ldots\omega_N)} q^{\#T(\omega_{m+1}\ldots\omega_N)} X(\omega_1 \ldots \omega_N)\\
&= \sum_{\omega_{n+1}\ldots\omega_N} p^{\#H(\omega_{n+1}\ldots\omega_N)} q^{\#T(\omega_{n+1}\ldots\omega_N)} X(\omega_1 \ldots \omega_N)\\
&= \mathbb{E}_n[X](\omega_1 \ldots \omega_n).
\end{aligned}
$$

roof of (iv):

$$
\begin{aligned}
&\mathbb{E}_n[X](\omega_1 \ldots \omega_n) \\
&= \sum_{\omega_{n+1}\ldots\omega_N} p^{\#H(\omega_{n+1}\ldots\omega_N)} q^{\#T(\omega_{n+1}\ldots\omega_N)} X(\omega_{n+1}\ldots\omega_N) \\
&= \sum_{\omega_1\ldots\omega_n} p^{\#H(\omega_1\ldots\omega_n)} q^{\#T(\omega_1\ldots\omega_n)} \\
&\qquad \cdot \sum_{\omega_{n+1}\ldots\omega_N} p^{\#H(\omega_{n+1}\ldots\omega_N)} q^{\#T(\omega_{n+1}\ldots\omega_N)} X(\omega_{n+1}\ldots\omega_N) \\
&= \sum_{\omega_1\ldots\omega_N} p^{\#H(\omega_1\ldots\omega_N)} q^{\#T(\omega_1\ldots\omega_N)} X(\omega_{n+1}\ldots\omega_N) \\
&= \mathbb{E}X.
\end{aligned}
$$

roof of (v): Let φ be a convex function, and denote by $\mathcal{L}$ the collection of l linear functions l which lie below φ, i.e., such that $l(y) \le \varphi(y)$ for all y. hen, as shown in the proof of Theorem 2.2.5,

$$\varphi(x) = \max_{l \in \mathcal{L}} l(x)$$

r all x.

For a random variable X and for all $\ell \in \mathcal{L}$, we have $\varphi(X) \ge \ell(X)$, and nsequently, $\mathbb{E}_n[\varphi(X)] \ge \mathbb{E}_n[\ell(X)]$. On the other hand, by property (i) (lin- rity), $\mathbb{E}_n[\ell(X)] = \ell(\mathbb{E}_n[X])$ so that $\mathbb{E}_n[\varphi(X)] \ge \ell(\mathbb{E}_n[X])$ for all $\ell \in \mathcal{L}$. herefore

$$\mathbb{E}_n[\varphi(X)] \ge \max_{\ell \in \mathcal{L}} \ell(\mathbb{E}_n[X]) = \varphi(\mathbb{E}_n[X]).$$

References

1. ARROW, K. & DEBREU, G. (1954) Existence of equilibrium for a competitive economy, *Econometrica* **22**, 265–290.
2. BENSOUSSAN, A. (1984) On the theory of option pricing, *Acta Appl. Math.* **2**, 139–158.
3. BLACK, F. (1976) The pricing of commodity contracts, *J. Fin. Econ.* **3**, 167–179.
4. BLACK, F., DERMAN, E., & TOY, W. (1990) A one-factor model of interest rates and its application to treasury bond options, *Fin. Anal. J.* **46**, 33–39.
5. BLACK, F. & SCHOLES, M. (1973) The pricing of options and corporate liabilities, *J. Polit. Econ.* **81**, 637–659.
6. BRACE, A., GATAREK, D., & MUSIELA, M. (1997) The market model of interest-rate dynamics, *Math. Fin.* **7**, 127–154.
7. COX, J. C. & HUANG, C. (1989) Optimal consumption and portfolio policies when asset prices follow a diffusion process, *J. Econ. Theory* **49**, 33–83.
8. COX, J. C. & HUANG, C. (1991) A variational problem arising in financial economics, *J. Math. Econ.* **20**, 465–487.
9. COX, J. C., INGERSOLL, J. E., & ROSS, S. (1981) The relation between forward prices and futures prices, *J. Fin. Econ.* **9**, 321–346.
10. COX, J. C., INGERSOLL, J. E., & ROSS, S. (1985) A theory of the term structure of interest rates, *Econometrica* **53**, 385–407.
11. COX, J. C., ROSS, S., & RUBINSTEIN, M. (1979) Option pricing: a simplified approach, *J. Fin. Econ.* **3**, 145–166.
12. COX, J. C., ROSS, S., & RUBINSTEIN, M. (1985) *Options Markets*, Prentice-Hall, Englewood Cliffs, NJ.
13. DOOB, J. (1942) *Stochastic Processes*, J. Wiley & Sons, New York.
14. DUFFIE, D. (1992) *Dynamic Asset Pricing Theory*, Princeton University Press, Princeton, NJ.
15. GEMAN, H., EL KAROUI, N., & ROCHET, J.-C. (1995) Changes of numéraire, changes of probability measure, and option pricing, *J. Appl. Prob.* **32**, 443–458.
16. HAKANSSON, N. (1970) Optimal investment and consumption strategies under risk for a class of utility functions, *Econometrica* **38**, 587–607.
17. HARRISON, J. M. & KREPS, D. M. (1979) Martingales and arbitrage in multiperiod security markets, *J. Econ. Theory* **20**, 381–408.
18. HARRISON, J. M. & PLISKA, S. R. (1981) Martingales and stochastic integrals in the theory of continuous trading, *Stochastic Processes Appl.* **11**, 215–260.

19. Heath, D. (1995) A continuous-time version of Kulldorf's result, preprint, Department of Mathematical Sciences, Carnegie Mellon University.
20. Heath, D., Jarrow, R., & Morton, A. (1992) Bond pricing and the term structure of interest rates: a new methodology for contingent claims valuation, *Econometrica* **60**, 77–105.
21. Heath, D., Jarrow, R., & Morton, A. (1996) Bond pricing and the term structure of interest rates: a discrete time approximation, *Fin. Quant. Anal.* **25**, 419–440.
22. Ho, T. & Lee, S. (1986) Term-structure movements and pricing interest rate contingent claims, *J. Fin.* **41**, 1011–1029.
23. Hull, J. & White, A. (1990) Pricing interest rate derivative securities, *Rev. Fin. Stud.* **3**, 573–592.
24. Jamshidian, F. (1997) LIBOR and swap market models and measures, *Fin. Stochastics* **1**, 261–291.
25. Jarrow, R. A. & Oldfield, G. S. (1981) Forward contracts and futures contracts, *J. Fin. Econ.* **9**, 373–382.
26. Karatzas, I. (1988) On the pricing of American options, *Appl. Math. Optim.* **17**, 37–60.
27. Karatzas, I., Lehoczky, J., & Shreve, S. (1987) Optimal portfolio and consumption decisions for a "small investor" on a finite horizon, *SIAM J. Control Optim.* **25**, 1557–1586.
28. Karatzas, I. & Shreve, S. (1998) *Methods of Mathematical Finance*, Springer, New York.
29. Kolmogorov, A. N. (1933) Grundbegriffe der Wahrscheinlichkeitsrechnung, *Ergeb. Math.* **2**, No. 3. Reprinted by Chelsea Publishing Company, New York, 1946. English translation: *Foundations of Probability Theory*, Chelsea Publishing Co., New York, 1950.
30. Kulldorf, M. (1993) Optimal control of a favorable game with a time-limit, *SIAM J. Control Optim.* **31**, 52–69.
31. Margrabe, W. (1978) A theory of forward and futures prices, preprint, Wharton School, University of Pennsylvania.
32. Merton, R. (1969) Lifetime portfolio selection under uncertainty: the continuous-time case, *Rev. Econ. Statist.* **51**, 247–257.
33. Merton, R. (1971) Optimum consumption and portfolio rules in a continuous-time model, *J. Econ. Theory* **3**, 373–413. Erratum: *ibid.* **6** (1973), 213–214.
34. Merton, R. (1973) Theory of rational option pricing, *Bell J. Econ. Manage. Sci.* **4**, 141–183.
35. Merton, R. (1973) An intertemporal capital asset pricing model, *Econometrica* **41**, 867–888.
36. Merton, R. (1990) *Continuous-Time Finance*, Basil Blackwell, Oxford and Cambridge.
37. Pliska, S. R. (1986) A stochastic calculus model of continuous trading: optimal portfolios, *Math. Oper. Res.* **11**, 371–382.
38. Sandmann, K. & Sondermann, D. (1993) A term-structure model for pricing interest-rate derivatives, *Rev. Futures Markets* **12**, 392–423.
39. Shiryaev, A. N. (1978) *Optimal Stopping Rules*, Springer, New York.
40. Shiryaev, A. N. (1999) *Essentials of Stochastic Finance: Facts, Models, Theory*, World Scientific, Singapore.

41. SHIRYAEV, A. N., KABANOV, YU. M., KRAMKOV, D. O., & MELNIKOV, A. V. (1995) Towards the theory of pricing options of both European and American types. I. Discrete time, *Theory Prob. Appl.* **39**, 14–60.
42. VASICEK, O. (1977) An equilibrium characterization of the term structure, *J. Fin. Econ.* **5**, 177–188.
43. VILLE, J. (1939) *Étude Critique de la Notion du Collectif*, Gauthier-Villars, Paris.

Index